A Practical Handbook on Sheep and Wool for the Farmer

by Geo. Jeffrey

with an introduction by Jackson Chambers

Self Reliance Books

Get more historic titles on animal and stock breeding, gardening and old fashioned skills by visiting us at:

http://selfreliancebooks.blogspot.com/

Introduction

I am pleased to present yet another practical title on breeding and raising livestock.

The work is in the Public Domain and is re-printed here in accordance with Federal Laws.

As with all reprinted books of this age that are intended to perfectly reproduce the original edition, considerable pains and effort had to be undertaken to correct fading and sometimes outright damage to existing proofs of this title. At times, this task is quite monumental, requiring an almost total "rebuilding" of some pages from digital proofs of multiple copies. Despite this, imperfections still sometimes exist in the final proof and may detract from the visual appearance of the text.

I hope you enjoy reading this book as much as I enjoyed making it available to readers again.

Jackson Chambers

DEDICATION.

TO

Albert Edwin Hamilton,

MEMBER OF THE BOARD OF EXAMINATION, SCHOOL OF
MINES AND INDUSTRIES. MEMBER OF THE PASTORAL
COMMITTEE, ROYAL AGRICULTURAL SOCIETY, AND
THE ONE WHO WAS MAINLY INSTRUMENTAL IN
STARTING THE WOOL CLASSES AT THE
SCHOOL OF MINES AND INDUSTRIES.

In grateful recognition of sincere friendship.

I DEDICATE THIS BOOK.

Preface.

During the course of my connection with the Rose-worthy Agricultural College, it was a pleasure to me to follow the work Professor Perkins was undertaking in connection with the sheep carrying capacity of our farms. I was very much impressed with the importance of the work he had in hand, but I must confess to having been staggered by the immense possibilities to our farming areas opened up by the publication in May, 1907, of his able report on the College flock for the year 1906, and, although I may not see eye to eye with the Professor on every point, it occurred to me that the report was altogether too valuable to be left to printing in the newspapers—valuable instructors as those organs are—or even to be recorded in a pamphlet or journal, because publications of these kinds are very apt to be laid aside and lost sight of. What seemed to me the proper method of placing the report permanently within the reach of all, was to publish it in book form, and at the same time to take advantage of the opportunity of supplementing it with other information respecting sheep and wool that had not been dealt with by the Professor.

When I mentioned the matter to Professor Perkins he readily agreed to the report being incorporated in a work such as I suggested, and generously offered to assist me by correcting proofs and in any other way he could. My Directors heartily approved of the project, and decided to issue 6,000 copies of the book to those interested in sheep farming.

I may add that I feel confident that in the near future the sheep industry will be a more important factor than ever in South Australia's welfare, and, as there are so many beginners in the business, I trust this little book will be of some service.

For many years I have been known in the State as a teacher dealing with matters relating to sheep and wool, and have already written a book on woolclassing, and, as my advice is frequently sought in connection with the care and handling of sheep, I trust that this will be accepted as ample justification for the present work.

I gratefully express my best thanks to Professor Perkins for his kindness and encouragement all through my work, and to Professor Angus and Mr. G. A. W. Pope for many valuable hints and the loan of photographic blocks, and to my respected colleague, Mr. George Dowling, and other members of the staff in the business with which I am connected for the deep and helpful interest they have displayed in the publication.

INDEX.

REFERENCE TO ADVERTISEMENTS.

Introduction.

That South Australia offers great advantages to the industrious man on the land cannot be denied, for, despite the cry about cheaper land elsewhere, it is difficult to imagine a more prosperous class of farmers than those inside what is known as Goyder's line of rainfall in South Australia. For stock raising and wool and fruit growing, this area is eminently suited. There is no doubt that during the last ten or fifteen years things have taken a wonderful turn for the better, largely owing to the use of artificial manures and the ready way in which our land has responded to these. This applies not only to the growing of cereal, fodder, and other crops, but is equally true with regard to stock raising, for it is a fact that since the advent of the artificial fertilisers, the carrying capacity of the land from their use has been increased in many cases threefold, and it is very pleasing to know that on much land known as second and third-class country, where scarcely a blade of grass grew, we now find numbers of sheep and cattle.

The establishment of the frozen meat industry has also largely assisted in bringing about this wonderful improvement. The good work done at the Agricultural College, Roseworthy, has certainly materially assisted in bringing about the better conditions existing to-day, and with the extension of agricultural education in the direction of establishing experimental farms throughout the State, in addition to the experimental work carried on at different centres under the Department of Agriculture, we can reasonably hope that the benefits will continue increasing in future years.

INTRODUCTION.

The School of Mines and Industries is also doing an excellent work in the teaching of Woolclassing and kindred subjects.

In what is known as the outside country there are signs of improvement, and we look forward with interest to the result of the newer methods of dry farming which are being carried out.

Land values in the country are higher than ever they have been in South Australia ; but it is just possible that when the extremely high rates now obtaining for practically every product begin to decline that land values also may fall slightly in sympathy. There is no doubt, however, that the fall, if it does come, can only be a temporary one, for, as our farmers become more experienced and as the influence of the various institutions referred to becomes more felt in South Australia, the production of the land must materially increase.

School of Mines Students at Work.

The Roseworthy Agricultural College Farm Flock in 1906-7.

Report by ARTHUR J. PERKINS, Principal Roseworthy Agricultural College.

———

It has not in the past been the practice to submit any special reports on live-stock operations in which the College farm may have been concerned, very largely, perhaps, because comparatively little had been done in this direction. It is, however, fairly evident to-day that the days when South Australian farmers could confine their attentions almost exclusively to the raising of grain and hay crops are fast disappearing ; with the steady and progressive rise in land values we must discover means tending to raise the average returns of our farms. The experience of other countries is available to show that this can only be realised by combining the rearing and grazing of live-stock with the raising of crops that can at all times command a ready market. I propose, therefore, issuing periodically, in addition to the ordinary crop reports, special reports having reference to live-stock operations likely to be of value to the farming community. The present report concerns mainly the College flock during the past season.

LIVE-STOCK AND PRESENT-DAY FARMING.

Whilst it is fairly clear that we must combine live-stock with our ordinary farming, the methods and practices best adapted to our altered conditions are not equally clear ;

indeed, we can only hope to reach finality in the matter by the experience of successive years.   That some modifications in our usual routine are imperative cannot, however, be denied ; but to what lengths are we to go ?   It is conceivable, for instance, to imagine a farm on which all crops raised are fed to some form or other of live-stock.   This would tend to maintain the land in the highest state of fertility, and represents, indeed, the ideal professedly aimed at in some countries. As a State, however, we are essentially exporters of wheat, and are likely to continue so for many years to come ; and in present circumstances, therefore, it appears to me the aim before us is twofold—(1) how to derive the greatest direct profit from our farm and live-stock, and (2) how to use them to the best advantage towards increasing the average yields of our staple export crops.

We have first to determine what form of live-stock will come nearest to fulfilling these ends ; and here let it be said that in a country of the dimensions of South Australia, with its great variations in climatic conditions, no general rule on the subject holding good for the whole State is conceivably possible.   And what I may have to relate in these reports I take to be applicable mainly to those portions of the Lower North in which conditions similar to our own obtain.   At the same time I trust that they may have their interest for those differently situated.   Our choice lies between sheep, pigs, and cattle, or some combination of the three ; but there are few who would care to dispute the fact that in the Lower North, at all events, sheep must constitute by far the most general form of farm live-stock.   Let me say that in his time Professor Lowrie never wearied of preaching sheep to farmers ; but it is only latterly that his advice appears to begin to be heeded to any degree.   It is certainly unnecessary to dwell upon the great adaptability of sheep to conditions obtaining on our Lower North farms : their great hardiness in times of scarcity, their ability to turn to best advantage the characteristically short herbage and dry feed of these districts, their comparative immunity from disease, their value in keeping weeds down, and supplementing cultivation, are too well known to need repeating. Whether in course of time cattle will succeed in edging them out of the better pastures of these regions is not a question that concerns us at present. I will merely state that in present circumstances sheep appear to be the form of live-stock best adapted to the conditions of the College farm.

Sheep-carrying Capacity of our Farms.

When a farm is to be stocked with sheep the first point for consideration will have reference to the number that can be kept to best advantage on the available area. Now, whilst I readily recognize that there is greater wisdom in understocking than in overstocking, I must protest against the commonly held opinion that rules that hold good for the stocking of station property must necessarily hold good for our farms. If this were so, beyond the small direct profit that can always be derived from a small flock of sheep, we should reap but little advantage from the presence of a flock on the farm. Let us remember that sheep are there, not only to show a direct profit balance of their own, but to help us in our cropping, to enrich our land, to keep down undesirable weeds, and to turn to advantage odds and ends that would otherwise be lost to us. And whilst I believe that even where general conditions are best known none of us is as yet in a position to state very definitely what number of sheep to a given area our farms are capable of carrying to advantage, I am nevertheless of opinion that this number is much in excess of what is generally believed to be the case. I think, too, that the farmer who wishes to combine sheep with his cropping must not limit their number to what he deems to be the carrying capacity of his available grazing paddocks at the worst time of the year. In other words, he must always hold himself in readiness to feed his sheep, should occasion arise ; he must raise fodder crops for their benefit, crops that will eventually pave the way for heavier wheat crops.

During the last three years I have been steadily testing the carrying capacity of the College farm with respect to sheep, and every year I have been able to increase the size of the flock, without in any way reducing the area under cultivation. Every year, in September, I have indicated the number of sheep at the time depastured on the College farm. It has been pointed out to me, however, that this did not give any indication of what we were able to carry throughout the year. To meet this objection I have kept a careful note of the number of sheep present at the beginning of each month ; the information is given below in tabular form :—

TABLE I.

1906	April 1	865 sheep
"	May 1	824 "
"	June 1	1,211 "
"	July 1	1,173 "
"	August 1	1,158 "
"	September 1	1,154 "
"	October 1	1,111 "
"	November 1	1,616 "
"	December 1	1,635 "
1907	January 1	1,550 "
"	February 1	1,464 "
"	March 1	1,374 "
"	April 1	1,105 "

Now, these sheep, in addition to some 80 head of cattle and over 300 pigs, were carried on a farm 1,430 acres in area. Let us see what portion of it was available for grazing purposes. In the first place, exclusive of the permanent Experiment Field, there were 556 acres under cereal and pea crops. The stubbles of this area did not become available until the month of February. We can, therefore, assume that one-sixth of this area, that is to say, about 93 acres, was available for grazing purposes during the course of the year. Another 122 acres are absorbed by the permanent Experiment Field, and are never open to grazing; eight acres under yards and buildings are similarly situated. Of fallows, there were 380 acres, which were progressively broken up during the course of the winter. We may assume that the whole area was available for grazing purposes to the end of July—that is, one-quarter for the whole year, or 95 acres.

In addition to this, 350 acres were reserved for purely grazing purposes, and distributed as follows :—Stubble catch-crops (rape, rye, clover), 94 acres; thousand-headed kale, 20 acres; lucerne, 2 acres; natural pasture, 234 acres; a total of 350 acres. Thus the total grazing area available for the year may be represented as follows :—Stubbles, 93 acres; pasture before fallowing, 95 acres; area exclusively grazed. 350 acres; a total of 538 acres.

Thus 538 acres carried successfully an average of 1,354 sheep, in addition to being grazed by 70 to 80 head of cattle and 250 to 300 pigs. The pigs and cattle were, of course, partly hand-fed. When it is remembered that in its natural state the land around the farm is very poor grazing land it will be realised how important an item sheep are capable of becoming on farms that are better situated.

Sheep-carrying Capacity of Individual Fields.

This question of the carrying capacity of our farm lands will, however, I believe, bear looking into more in detail. I propose, therefore, reviewing the results from typical fields, the records of which have been carefully kept. One field, known as No. 16, carried a self-sown crop of barley and natural herbage; another, No. 7A, was sown to thousand-headed kale; and the third, No. 3, was sown to a catch-crop of rape, mustard, and rye.

Field No. 16.

This field, 60 acres in area, is probably the poorest on the College Farm. It was covered originally with low mallee scrub, and consists largely of light limestone land, and is more or less covered with stones on the surface. About 6 to 7 acres of this field are nothing more than a drifting native-pine sand-dune. On the whole, it will be seen that in ordinary circumstances it would be looked upon as exceedingly poor grazing land. This field carried in 1905-6 a crop of barley dressed with 2 cwt. of superphosphate. Sheep were turned into the stubbles in February and March, 1906. We shall take up the history of the paddock from April 1, 1906, and, in view of the importance of the subject, I trust I shall be excused for giving what might otherwise appear to be wearisome details.

Table II.

Dates.	Number of days.	Sheep grazing on 60 acres.
1906 April 1 to 18	18	780
" April 19 to May 23	36	462
" May 25 to June 25	33	360
" July 16 to 24	8	502
" August 12 to 13	2	301

Dates.	Number of Days.	Sheep grazing on 60 acres.
1906 August 31 to Sept. 6 ..	7	271
" September 7 to 17	10	281
" October 4 to 7	4	192
" November 3 to 5	3	758
" November 6 to 12	7	555
" November 13 to 20	8	142
" November 21 to Dec. 2	12	192
" December 3 to 16	14	292
" December 17 to Jan. 4	19	249
1907 January 5 to 14	10	215
" January 15 to 22	8	294
" January 23 to April 1 ..	68	281

Towards the latter part of March the sheep were provided with a little cocky chaff and molasses ; but they were not otherwise hand-fed. The figures in the above table work out to 260 sheep to the field for 365 days, or roughly, four and one-third sheep to the acre for the year ; and it will be noticed that it has been grazed more or less continuously throughout the year.

Horses and cattle were also put to graze in this field for several days during the course of the year. On the assumption that one horse or cow represents the grazing capacity of 5 sheep, this does not represent more than an additional 4 sheep to the paddock for 365 days. I have referred to this fact merely to show that, notwithstanding the large number of sheep present, the field continued to show sufficient length of feed to permit of our turning into it from time to time a few head of cattle or horses.

I am inclined to look upon the results obtained in this field as remarkable. In its natural state the soil is very poor, and its present high grazing capacity must be set down to the use of heavy dressings of phosphates on preceding cereal crops.

Field No. 7A.

In the preceding field we have dealt with feed growing spontaneously on the stubbles of a cereal crop ; in the present one we have a crop sown specially to be fed down by sheep. It is 20 acres in area. The soil in this field is good medium loam on clay bottom, and in point of view of natural fertility a striking contrast to the preceding one. In 1905-6 it carried a mixture of oats, wheat, and vetches, sown for ensilage ; it cut $8\frac{1}{2}$ tons of green stuff to the acre. The stubbles were broken up with an ordinary 3-furrow stump-jump plough, from March 13 to 23, and rolled and cultivated from March 30 to April 3. The thousand-headed kale seed was drilled in on April 3 and 4, at the rate of 1 lb. to the acre, mixed with $\frac{1}{2}$-cwt. of bonedust, in rows 32 inches apart. The field was lightly harrowed after seeding. The early season was not altogether favorable to the growth of the kale ; April and May were exceptionally dry, and the crop made but poor progress in these months. Later on, however, it shot away, and presented very fine growth in the early spring. Sheep were not placed in this field until September 6, so that it was fed down only during seven months of the twelve. Unquestionably, we might have started feeding it down earlier in the season in sections. We should not then have lost quite as much from treading down ; and had this been done it is probable that the field would have carried a greater number of stock.

The details of sheep fed are shown below in Table iii.

TABLE III.

Dates.	Number of days.	Sheep on 20 acres.
1906 September 6 and 7	2	270
" September 14 and 15 ..	2	270
" September 16	1	550
" September 17 to 26 ..	10	776
" October 8 and 9	2	773
" October 14 to 20	7	673
" October 24 to 25	2	673
" October 27 and 28	2	758
" October 29 to Nov. 5 ..	8	63
" November 6 and 7	2	187
" November 8 to 10	3	63
" November 11 to 12	2	169
" November 13 to 15	3	308
" November 16 to 18	3	299
" November 19 to 20	2	292
" November 21	1	290
" November 22	1	286
" November 23	1	281
" November 24 to 25	2	279
" November 26	1	270
" November 27 to Dec. 20	24	207
" December 21 to Jan. 18	29	18
1907 February 27	1	993

The above figures represent 84 sheep on 20 acres for 365 days, or roughly four and one-fifth sheep to the acre for the year. When we consider the fact that the paddock was not entered until September, I think that it will be agreed that the figures are fairly satisfactory. Indeed, the great advantage of kale in a dry climate is that without the aid of irrigation it supplies an abundance of succulent feed, on which lambs can be weaned to great advantage.   I am of opinion, however, that in more favorable seasons the carrying capacity of a field of kale is even greater than that shown above.

Field No. 3.

Lastly we come to a field in which the stubbles of a cereal crop were lightly cultivated, and made to carry a fodder catch-crop.   It is 43 acres in area.   In quality this field is intermediary with Field No. 7A and Field No. 16, that is a fair medium loam, running here and there into loose limestone soil.

The stubbles of a hayfield were cultivated over with a chisel-tined cultivator from March 28 to April 5, the ground was rolled from April 4 to 6, and sown from April 5 to 9, the seed being harrowed in.   The following mixture was used for seeding purposes :—Dwarf Essex Rape, 6 lb. per acre ; Rye, 30 lb. per acre ; White Mustard, 3-20 lb. per acre. The 1906-7 season was not altogether favorable to crops of this sort, with the result that the grazing capacity of this field was not as high as might have been anticipated.   I append below in Table iv. details of the sheep depastured in this field :—

TABLE IV.

Dates.	Number of days.	Sheep on 43 acres.
1906 June 18 to 26	8	279
" June 26 to 30	5	301
" July 12 to 27	16	321
" August 14 to 31	18	283
" September 1 to 16	16	281
" September 18	1	271
" September 28 to 30	3	775
" October 1 to 8	8	773

	Dates.	Number of days.	Sheep on 43 acres.
1906	October 9 to 10	2	673
"	October 28 to Nov. 1 ..	5	758
"	November 17 to 22	6	63
"	November 23 to 26	4	188
"	November 27 to Dec. 8	12	127
"	December 9 to Jan. 17	40	190
1907	January 18 to 30	13	187
"	February 28 to March 22	23	36

The above figures work out roughly to about three sheep to the acre for 365 days ; and it will be noticed that from mid-June the field carried stock fairly continuously throughout the year. From August to the end of March this field carried, for fairly long periods, both cattle and horses. From data kept, the number of large stock carried was equivalent to half sheep per acre for 365 days. We may, therefore, take the grazing capacity of the field to have been represented by $3\frac{1}{2}$ sheep to the acre for the year. What has been said for the kale-field applies to even greater extent to the rape crop : a favorable autumn is essential to the success of this crop, and in more favorable seasons far better results may be anticipated.

RELATIVE VALUE FOR GRAZING OF THE THREE FIELDS.

Although I have given in detail the grazing capacity of three typical fields, under different methods of treatment, the results cannot be looked upon as finally settling the respective merits of these different practices. The season in a sense was in every way favorable to the self-sown barley that sprang up on the stubbles of Field No. 16 ; heavy rains fell towards the end of March, 1906, and brought on rapid germination of the grain that lay scattered over the surface ; and throughout a mild autumn and winter growth was practically continuous. On the other hand, the season was undoubtedly unfavorable to the later-sown rape and kale. The March rains were taken advantage of to break up the ground. By the time seeding had been completed, however, the soil had dried off considerably, and, unfortunately, no further rains of any importance fell before mid-June. Indeed, whatever the results, it is evident that each treatment has its advantages, and may be recommended : barley for early autumn and winter feed, rape for winter feed, and kale for spring and early summer feed.

An additional word or two on the value of barley for grazing purposes may not, however, prove out of place.

Its value in this direction is, no doubt, well known. We are all aware of the advantages of feeding down rank barley crops ; we know also its value for early green feed. I question, however, whether barley is made use of sufficiently freely on light cultivation for purely grazing purposes. For this object, the following practice appears to me the best :— Let a suitable hay-stubble paddock be selected ; tear it open with a chisel-tined cultivator, or multi-furrow skim-plough, as soon after harvest as possible, say some time in February. Roll the ground and broadcast over it from one to two bushels of barley, and harrow in the seed lightly. It is essential that the grain be buried very shallow, so that on the first autumn showers it be in a position to spring away, and furnish an abundant supply of succulent feed. Rape will follow it, and finally will come kale, on better cultivation, and at a time when other crops have withered away and died.

Some Conclusions.

Now, although I have indicated what we have done here during the past season, I am very far from imagining that the results of one year can furnish general rules for others to follow. Much depends on the seasons, but more, perhaps, on the state of preparedness in which one may find oneself. It is more than useless to lay oneself out to carry a large flock unless one be duly prepared for all eventualities; and whatever may be the case on a station, on a farm sheep should never be allowed to lose condition. And to achieve this result one must always be prepared to make good any shortage of natural herbage. Again, a farmer who attempts to keep a large flock, with paddocks 200 to 300 acres in area, courts failure. In my humble opinion, on a farm in which sheep are extensively kept, no field should exceed 50 acres in area, and many of them should be smaller. The best practice, in my opinion, is to stock heavily small paddocks, and keep changing from time to time, so as to enable the feed to recover and sweeten. Briefly, in the season 1906-7, we have carried, without hand-feeding to any appreciable extent, nearly a sheep to the acre on a farm of 1,430 acres, in addition to supplying grazing-ground to 80 head of cattle and 40 horses. Out of this area 122 acres were devoted to purely experimental work, and were not, therefore, available for grazing purposes ; whilst 556 acres were under cereal crops, the stubbles of which were only available for the two months of the period under consideration ; and, finally, 380 acres were treated as bare fallow, and were not available after July. Similar records are being kept during the present season, and I trust next year to have an equally favorable tale to unfold.

THE FARM SHEEP FOR LOWER NORTH CONDITIONS.

Next arises the question of the type of sheep best adapted to general farm conditions, and here again we are confronted with a great diversity of opinion, arising partly from mere personal predilection, and partly because uniformity cannot be expected where surrounding conditions are frequently so diverse.   Again, I can only speak for the district with which I am connected, and I trust that in analysing the College Farm experiences I shall be able to leave in the background questions of personal predilection.   On one point I think I can assume, in the Lower North at all events, that we are all agreed, namely, that the farmer who keeps sheep must sacrifice almost everything to the lamb ; in other words, of two types of sheep equally well adapted to the district and general farm conditions, the sheep able to rear the earlier and finer lamb is likely to prove the more profitable.   I cannot, of course, say that we have tested here every possible type of sheep ; but we have tested several, and as we have found them, let me show them.

And, first, let me say that we find the home-bred sheep superior to the purchased station sheep.   I am not now speaking of quality, but merely of adaptability to farm conditions.   It is many months before the station sheep gets accustomed to its new surroundings ; the smaller paddocks, frequent fences, general turmoil, all tend to disquiet it, with the result that it does not thrive.   Nevertheless, if we are to rear our ewes on the farm, portion of our flock must go dry every year in most cases ; and if we breed from 2-tooths, 1-5th, on the assumption that one-quarter of the ewes are replaced every year.   Whether, then, it will prove more profitable in a flock of 1,250 ewes to carry 250 hoggets yearly instead of the full complement of breeding ewes, is a matter for personal consideration.   Personally, I must decide in favor of the home-bred ewe and the latter practice.

Next, are Merino ewes, or half-bred ewes, to mother our

export lambs ? And if half-bred ewes, what crosses are to be preferred ? This, no doubt, is a knotty question, which cannot be solved to the liking of all. My old friend, Mr. Jeffrey is, I know, strongly in favor of the large-framed Northern Merino ewe. His main argument in their favor is, I believe, that whilst the latter can always readily be purchased in the open market, suitable half-bred ewes are always difficult to secure, and at times not obtainable for love or money. If the farmer breeds his own ewes, this difficulty disappears, and we have then, in favor of the half-bred ewe, in our experience at all events, that she is quieter and more thrifty, retains her condition more readily, is a better mother, and generally rears a more profitable lamb. The farm sheep of the Lower North, will, in my opinion. in the course of time be some form or other of half-bred sheep, bred from a large-framed Merino ewe.

Against the half-bred sheep it is frequently urged that it has but little respect for the ordinary fence, and is hence a source of danger to the farmer's crops. This, no doubt, is true. I hold, however, that on a farm, where sheep are being kept, the ordinary station 6-wire fence, unless reinforced by anchor droppers, is of very little value. With a fence of this kind for sole protection, the crops become a temptation that no farm-bred sheep can withstand. Plain fences that are to be kept sheep-proof must, on ordinary farms, carry 8 wires.

Then, again, it is urged, and with reason in most cases. that the fleece of the Merino ewe is more valuable than that of the half-bred ewe. Latterly, however, this has not always proved to be the case. With us certain crosses will carry a fleece, equal in weight, if not superior, to that of the Merino. and with high prices prevailing for crossbred wool the half-bred ewe has frequently proved the more profitable of the two, both as mother and fleece-carrier. We are not in a position, from the College records, to compare all possible crosses one with the other; we can only compare those that I have thought best adapted to our conditions. The longwool crosses—Lincoln Leicester or Romney Marsh—are absent. Rightly or wrongly, I do not look upon them on Lower North farms as equal to short-woolled crosses—Shropshire, South Down, or Dorset-Horn. I append below, in Table v., our average fleeces for the past three years. I do not hold up these figures as in themselves good. With us the lamb comes first, the wool second. Our lambs have always sold at the top of the market; our fleeces might easily be improved upon. Nor should it be forgotten that our shearers are, for the most part, novices, and apt to leave some of the wool on the backs of the sheep. The various types of sheep, however, have undergone fairly uniform treatment, and the results, therefore, however questionable as a standard for individual types, may fairly be taken to be comparable one with the other.

It should be added that on the whole the flock contains a greater number of aged sheep than will be the case in future years; I had to build up the flock from small numbers, and, although I have on two occasions purchased drafts of outside ewes, I have generally endeavoured to keep our own ewes as long as possible. This will have the effect of reducing the all-round average weight of fleece. When the full complement of breeding ewes has been attained, the flock will be renewed automatically every fourth year, with the result that the average weight will rise in proportion. We shall also be better situated to cull out unsuitable ewes than has been the case in the past.

TABLE V.

Showing Average Weight of College Ewe Fleeces, 1904-6.

	1904.		1905.		1906.	
	Number of Ewes.	Average Weight of Fleece.	Number of Ewes.	Average Weight of Fleece.	Number of Ewes.	Average Weight of Fleece.
		lb. oz.		lb. oz.		lb. oz.
Merino	379	8 10	351	8 14	284	9 14
Merino-Shropshire .	5	9 5	25	8 7	74	9 2
Merino-South Down	—	— —	10	7 6	54	8 0
Merino-Dorset Horn	71	7 9	96	6 13	81	6 15

Thus, on the whole, Merinos on the farm show slightly heavier fleeces than any of the shortwool crosses ; and for the latter, in order of weight of fleeces, we have first the half-bred Shropshires, then half-bred South Downs, and finally the half-bred Dorsets.

To the advantage of the light-fleeced Merino Dorset-Horn ewes, it may be said that they are excellent mothers and probably rear better lambs than any other sheep. With them twins are the rule rather than the exception, and in most cases they are able to rear them better than Merino ewes their single lambs.

The Merino South Downs are rather small, shapely sheep, and in their way excellent nurses. On the whole, however, I am inclined to think that they are not quite as profitable as the Merino Shropshire or Merino Dorset-Horn ewes. It is, in fact, well-bred ewes of these two types that I favor most for our conditions.

What Rams Should Sire Farm Flocks ?

Given the ewes, what rams should we make use of when the main object in view is the rearing of prime early lambs for export purposes ? As with the ewes, so with the rams, there appears to be no general unanimity of opinion. When we have said that the ram must be of an improved early-maturing British breed. we have covered all the ground concerning which there can be no two opinions ; but, there-after, it is as many opinions, as there are men, or, rather, breeders. And the fault lies very largely with the lamb-buyers ; not that we expect them to tell us what cross to select, but they might be expected to show some consistency in their views ; or, at all events, their officially-expressed views should be in accord with their purchases. When we find the freezer described by them as a 36-lb. lamb, and everything in excess as unsuitable for the market, we know where we are. When, however, on the other side, we find the heavier lamb, which can more readily be raised in the Lower North, fetching higher prices than the classic 36-pounder, we are puzzled, to say the least of it ; but usually we follow the market.

Early maturity and development is so important a feature in the rearing of lambs that we can readily sacrifice to it at times other points of lesser moment. In this direction the half-bred Dorset-Horn lamb appears. in our experience, generally to run away from any other cross. It may not have that property of laying on fat to excess. so characteristic of Lincoln and Leicester ; the lamb, nevertheless, develops exceedingly rapidly, and is ripe sooner than any other lamb we have experience of. There, however, its advantages end. It is not, on the whole, a shapely lamb, strong in the fore-quarter but weak behind, and is apt to be coarse, particularly if castrated at all late.

The half-bred Shropshire lamb, probably the most common in this State, does not mature as rapidly as the preceding one. It is, probably, when well-bred, slightly better proportioned, but still inclined to be heavy and coarse. The half-bred South Down lamb, particularly if mothered by the half-bred ewe, gives the ideal of perfect symmetry and proportion. True, it is generally lighter than the former two ; but what of that, if lamb-buyers really mean that 36 lb. is what they are aiming at ?

Although any one of the above three crosses gives splendid marketable lambs, I am inclined to favor the South Down ram on a half-bred Shropshire or Dorset-Horn ewe ; and in this manner may not the conflicting views of all be met ? When Professor Lowrie visited us last year, he advocated very strongly the claims of the Leicester. I do not wish to run counter to the views of such an authority, even for our own district. Let me say, however, that in a country where lambs must develop rapidly in the wet winter months, and be prime by September, I doubt whether a lamb with a long staple is likely to better the short-staple crosses of the Down breeds. I may, of course, be wrong. Many, no doubt, will put Professor Lowrie's advice to practical test, and their experience will certainly be awaited with interest.

Some Tests of Various Types of Crossbred Lambs made in 1906.

It has at different times been proposed that the relative value of rams of the different breeds, as getters of early lambs, be tested by carefully following out the progress of their progeny. In principle the proposal sounds easy enough ; n practice it is exceedingly difficult to carry out. In the first place, individually, one ram may be better than another of a different breed, and yet on his performances we would not be justified in condemning a well-established breed with which he had been matched. Again, who can guarantee perfect evenness of quality and temperament in ewes put to rams of the different breeds ? If the number of lambs under observation is small, we can guarantee general uniformity of treatment ; but the inherent peculiarities of individuals would be too pronounced. and thus rob the test of much of its value. If, on the other hand, notwithstanding the tediousness of the work involved, we place under observation large numbers of lambs—the whole flock, for instance—to secure equality of treatment is quite impossible. No two paddocks, for instance, offer equal grazing-grounds, either as to quality of the feed or the shelter afforded ; and upon these questions, of course, depends the development of the lamb.

Last season, however, I determined to try and find out what could be done in this direction. I decided that what we wanted to find out was how the different crosses behaved under ordinary farm conditions. To secure this result, I put all the lambs on the farm under observation. This involved tediously long individual weighings and measurements, and, moreover, the lambs had to take their chance as to general treatment. On the whole, however, I believe that the results I am about to give are more reliable than those that might have been secured from the weights and measurements of a few picked individuals under uniform treatment. I recognise that it is hardly fair to generalise from the data of a single season, and I propose, therefore, repeating the work several seasons in succession. The results referred to are tabulated below in Tables vi., vii., viii., and ix.

Sheep on Agricultural College Farm, Roseworthy.

TABLE VI.

Showing Growth and Development of 82 Lambs by Dorset-Horn Rams out of Merino Ewes.

Dates when Lambs Dropped	First Weighing, June 25					Second Weighing, August 3				Average Increase in Weight		Third Weighing, September 10					Average Increase in Weight (38 Days)		Average Increase in Weight (77 Days)		Fourth Weighing, October 25					Average Increase in Weight (83 Days)		Average Increase in Weight (122 Days)	
	Numbers	Length	Height	Girth	Weight	Length	Height	Girth	Weight	For 39 Days	Per Day	Numbers	Length	Height	Girth	Weight	For 38 Days	Per Day	For 77 Days	Per Day	Numbers	Length	Height	Girth	Weight	For 83 Days	Per Day	For 122 Days	Per Day
Wethers.																													
April 23 to 30	6	21.83	21.33	26.42	45.58	24.75	23.75	29.00	72.17	26.59	0.68	4	28.13	26.38	32.63	91.00	20.50	0.56	47.50	0.62	2	28.50	29.00	35.33	113.30	37.86	0.46	63.36	0.52
May 1 to 15	26	20.25	20.02	24.69	39.06	23.58	23.44	28.63	66.46	27.40	0.70	14	27.11	26.11	32.75	90.39	21.36	0.54	49.47	0.64	10	27.85	28.00	35.15	101.26	39.06	0.47	65.16	0.53
May 16 to 31	7	18.79	19.00	22.71	31.64	22.57	23.07	27.21	59.43	27.79	0.71	1	26.50	27.00	31.50	88.00	22.00	0.58	53.50	0.69	5	27.70	27.40	35.40	103.52	45.72	0.53	73.42	0.60
Total	39	20.23	20.04	24.60	38.69	23.58	23.42	28.44	66.04	27.36	0.70	19	27.29	26.21	32.66	90.39	21.21	0.56	49.27	0.64	17	27.88	27.97	35.25	103.35	40.88	0.49	67.38	0.55
Ewes.																													
April 23 to 30	1	20.00	18.00	26.50	37.06	22.00	23.00	25.50	60.06	23.00	0.59										1	28.00	27.50	34.50	88.00	28.00	0.34	51.00	0.42
May 1 to 15	27	19.74	18.63	25.57	34.05	22.69	22.41	27.99	60.41	26.36	0.68										25	27.24	28.00	35.69	93.22	32.81	0.40	59.56	0.49
May 16 to 31	14	18.43	17.86	23.68	37.79	22.32	22.00	26.57	54.61	26.82	0.69										14	26.82	26.98	34.21	90.68	32.96	0.43	52.96	0.52
After May 31	3	16.00	14.50	19.50	18.00	20.00	20.00	24.50	42.00	24.00	0.62										1	26.00	29.00	33.00	76.00	34.00	0.41	58.00	0.48
Total	43	19.23	18.25	24.83	31.71	22.02	21.77	26.87	58.10	26.39	0.68										41	27.07	27.65	35.09	91.81	33.82	0.41	60.21	0.49
GENERAL AVERAGE OF WETHER AND EWE LAMBS.																													
Total	82	19.71	19.11	24.72	35.68	22.76	22.55	27.62	61.63	26.60	0.68										58	27.31	27.74	35.14	95.19	35.89	0.46	62.31	0.51

Notes on Table VI.

This table shows the development of 82 lambs by Dorset-Horn rams out of Merino ewes, between June 25 and October 25. I append a few general remarks on this table.

The dimensions of the lambs taken in all instances were length, height, and girth. Length represents the distance between the base of the neck and the root of the tail, the height is taken at the point of the shoulder, and the girth immediately behind the shoulder. All dimensions are shown in inches and their decimals.

From June 25 to August 3, wether lambs of this cross increased in weight at the rate of 7-10th of a lb. per diem, and ewe lambs slightly less. For wethers of this cross the highest individual increase during this period was from $44\frac{1}{2}$ lb. to $79\frac{1}{2}$ lb.—that is to say, at the rate of 9-10th of a lb. per diem ; for ewes, from $20\frac{1}{2}$ lb. to 54 lb.—that is to say, at the rate of about 17-20th of a lb. per diem. For wethers, the lowest increase was from $35\frac{1}{2}$ lb. to 52 lb.—that is to say, at the rate, approximately, of 2-5th of a lb. per diem ; and for ewes, from $30\frac{1}{2}$ lb. to 51 lb.—that is to say, at the rate, approximately, of $\frac{1}{2}$-lb. per diem.

From August to the end of October, the average rate of increase in weight was reduced to $\frac{1}{2}$-lb. per diem for wether lambs, and 2-5th of a lb. for ewe lambs.

I regret to say that, so far as the wether lambs are concerned, the figures of the third and fourth weighings are not strictly comparable. During the period covered by these two weighings, lambs that were ripe were sold, whilst the less forward remained on. Unfortunately, on September 10, the date of the third weighing, we found ourselves too busy with

TABLE VII.

Showing Growth and Development of 42 Lambs by South Down Rams out of Merino Ewes.

Date when Lambs Dropped.	First Weighing. June 25.					Second Weighing. August 3.				Avg. Increase For 39 Days	Per Day	Third Weighing. September 10.					Avg. Increase For 38 Days	Per Day	Avg. Increase For 77 Days	Per Day	Fourth Weighing. October 25.					Avg. Increase For 83 Days	Per Day	Avg. Increase For 122 Days	Per Day
	Numbers	Length	Height	Girth	Weight	Length	Height	Girth	Weight	lb.	lb.	Numbers	Length	Height	Girth	Weight	lb.	lb.	lb.	lb.	Numbers	Length	Height	Girth	Weight	lb.	lb.	lb.	lb.
WETHERS—																													
April 23 to 30	1	20.50	19.00	24.50	39.50	20.50	23.00	28.00	60.00	20.50	0.53										1	28.00	29.00	35.00	88.00	28.00	0.34	48.50	0.40
May 1 to 15	12	19.62	19.79	23.92	36.00	22.88	22.63	27.08	57.67	21.67	0.56	2	26.25	25.50	32.75	90.50	24.50	0.65	46.00	0.60	10	26.55	25.70	33.95	91.00	35.60	0.43	57.30	0.47
May 15 to 31	8	18.00	18.50	21.44	27.25	20.69	21.63	25.50	50.88	23.63	0.61	1	27.00	25.00	33.00	85.00	21.00	0.55	48.50	0.63	7	25.14	25.79	33.64	83.93	34.93	0.42	58.00	0.48
After May 31	1	16.00	17.00	20.50	20.50	20.00	21.00	24.50	45.00	24.50	0.63										1	25.00	24.00	33.50	82.00	37.00	0.45	61.50	0.50
Total	22	18.89	19.16	22.89	32.27	21.91	22.20	26.43	54.73	22.46	0.58	3	26.50	25.33	32.83	88.67	23.33	0.61	46.83	0.61	19	26.03	25.82	33.88	88.09	35.03	0.43	57.26	0.47
EWES—																													
April 23 to 30	2	20.50	19.50	25.50	40.00	22.50	23.25	28.00	55.50	15.50	0.40										2	25.50	27.00	33.75	91.00	30.50	0.37	55.00	0.53
May 1 to 15	16	19.41	19.25	23.91	33.81	22.18	21.84	27.08	54.50	20.69	0.53										14	26.86	26.36	34.36	83.14	28.64	0.35	49.50	0.41
May 16 to 31	1	19.00	19.00	22.50	28.50	22.00	23.00	25.50	54.50	26.00	0.67										1	26.00	27.00	33.50	90.00	35.50	0.43	61.50	0.50
After May 31	1	15.50	15.00	18.00	18.50	18.00	20.00	24.50	41.50	23.00	0.59										1	26.00	25.50	33.00	75.00	33.50	0.40	56.50	0.46
Total	20	19.30	19.05	23.70	33.40	21.97	21.95	26.43	53.95	20.55	0.53										18	26.67	26.42	34.17	83.39	29.50	0.36	51.03	0.42
GENERAL AVERAGE OF WETHER AND EWE LAMBS. Total	42	19.08	19.12	23.27	32.81	21.94	22.08	26.57	54.36	21.55	0.55										37	26.31	26.11	34.01	86.07	32.34	0.39	54.26	0.44

other matters to weigh all the lambs, with the result that only those that left the farm were weighed. These remarks do not apply to the half-bred ewe lambs, all of which were retained for breeding purposes.

It follows, from what has been said above, that the averages of the fourth weighing, taken in October, are lower than would have been the case had not the best wether lambs been sold earlier in the season.

Finally, it should be remembered that the figures given represent the average of a whole flock, small and large, weak and strong, all included. I do not, therefore, bring forward these results as the best obtainable under favorable conditions; I look upon them. however, as fairly creditable to ordinary farm conditions. That the lambs were good may be judged from the fact that early in September we sold them all forward for 15s. a head on the farm, and that all wether lambs had left the farm by the end of October.

Notes on Table VII.

This table sets out the development of 42 lambs by South Down rams out of Merino ewes, from June 25 to October 25. It will be noticed that, during the first period—that is to say, from June 25 to August 3—these lambs gained less in weight than the half-bred Dorset-Horn lambs. The wethers increased at the rate of slightly less than 3-5th of a lb. per diem, and the ewe lambs slightly over half-lb. per diem. During these 39 days the half-bred Dorset lambs gained an individual average of 5 lb. 2½ oz. more than the half-bred South Down lambs. The maximum increase for wether lambs was from 39½ lb. to 67 lb.—that is to say, at the rate of 7-10th of a lb. per diem; and the minimum increase 36½ lb. to 54½ lb.—that is to say, at the rate of 9-20th of a lb. per diem.

TABLE VIII.

Showing Growth and Development of 138 Lambs by Shropshire Rams on Merino Ewes.

Days when Lambs Dropped	First Weighing					Second Weighing				Average Increase in Weight		Third Weighing					Average Increase in Weight		Average Increase in Weight		Fourth Weighing					Average Increase in Weight		Average Increase in Weight	
	Numbers	Length	Height	Girth	Weight	Length	Height	Girth	Weight	For 40 Days	Per Day	Numbers	Length	Height	Girth	Weight	For 55 Days	Per Day	For 95 Day	Per Day	Numbers	Length	Height	Girth	Weight	For 71 Days	Per Day	For 111 Days	Per Day
		in.	in.	in.	lb.	in.	in.	in.	lb.	lb.	lb.		in.	in.	in.	lb.	lb.	lb.	lb.	lb.		in.	in.	in.	lb.	lb.	lb.	lb.	lb.
WETHERS —																													
April 23 to 30	6	21.25	21.25	27.42	46.33	24.50	24.17	28.92	65.67	19.34	0.48	5	26.50	25.20	33.39	86.80	19.40	0.35	38.20	0.41	1	27.00	25.50	34.00	84.00	17.00	0.24	38.00	0.32
May 1 to 15	16	20.58	20.91	25.51	41.32	23.09	23.80	27.64	59.80	18.48	0.46	12	26.34	25.34	33.34	86.82	23.27	0.42	43.41	0.46	24	26.91	26.04	33.58	82.17	24.69	0.35	41.86	0.38
May 16 to 31	16	19.15	20.05	23.85	35.80	23.05	23.10	26.10	53.25	17.45	0.44	3	26.17	25.83	32.00	80.33	28.17	0.51	47.50	0.50	6	26.50	26.42	31.58	73.67	25.33	0.36	41.58	0.37
After May 31	1	20.00	21.00	28.50	33.00	22.50	23.00	28.50	46.00	13.00	0.33										1	24.00	28.00	32.00	69.00	23.00	0.32	39.00	0.35
Total	63	20.40	20.82	25.40	40.79	23.86	23.71	27.50	59.10	18.31	0.46	30	26.35	25.37	33.29	87.10	23.12	0.42	43.05	0.45	32	26.77	26.80	33.19	80.22	24.52	0.35	41.51	0.37
EWES —																													
April 23 to 30	6	22.33	20.83	26.54	42.25	23.67	23.50	27.67	60.67	16.42	0.41										6	26.00	25.75	33.92	78.50	17.17	0.24	33.59	0.30
May 1 to 15	61	20.47	20.84	24.83	38.31	23.09	22.69	26.80	54.66	16.35	0.41										61	26.39	26.27	32.73	78.80	23.03	0.34	40.07	0.36
May 16 to 31	7	19.14	19.14	22.50	30.00	21.46	21.43	25.57	49.14	19.14	0.48										7	26.43	26.28	31.93	74.07	24.93	0.35	44.07	0.40
After May 31	1	18.00	19.00	22.00	29.00	21.00	21.00	25.00	41.00	12.00	0.30										1	28.00	27.00	33.00	77.00	36.00	0.51	48.00	0.43
Total	75	20.37	20.25	24.71	37.89	22.91	22.61	26.73	54.43	16.55	0.41										75	26.37	26.24	32.75	78.31	23.84	0.33	40.19	0.36
GENERAL AVERAGE OF WETHER AND EWE LAMBS —																													
Total	138	20.38	20.51	25.02	39.21	23.34	23.12	27.08	56.57	17.36	0.43										107	26.49	26.41	32.88	78.88	23.90	0.34	40.59	0.36

With later weighings we have the same difficulty as with the preceding table—that is to say. some of the earlier wether lambs were sold at a time when we were not in a position to weigh the whole flock. As in the preceding case, however, all ewe lambs were retained for breeding purposes, and all figures concerning them are strictly comparable. It may be noted in passing that whilst on September 10 one-half of the half-bred Dorset lambs were saleable, not more than a quarter of the half-bred South Downs were equally forward.

Notes on Table VIII.

This table shows the development of 138 lambs by Shropshire rams out of Merino ewes, between July 6 and October 25. During the first period, corresponding to that of the first two lots, the half-bred Shropshire lambs gained less than either the half-bred Dorset-Horn or the half-bred South Down lambs. Reducing the figures to an interval of 40 days, whilst the half-bred Shropshire lambs were gaining $17\frac{1}{4}$ lb., the half-bred Dorsets gained $27\frac{1}{4}$ lb., and the half-bred South Downs over 22 lb. During this first period the wether lambs increased at the rate of 9-20th of a lb. per diem. with a maximum of 57 lb. to 81 lb.—that is to say, at the rate of 3-5th of a lb. per diem ; and a minimum of 33 lb. to 46 lb.—that is to say, at the rate of $\frac{1}{3}$ of a lb. per diem. During the same period, ewe lambs increased at the rate of 2-5th of a lb. per diem, with a maximum of 25 lb. to 48 lb.—that is to say, at the rate of slightly under 3-5th of a lb. per diem ; and a minimum of $37\frac{1}{2}$ lb. to 47 lb.—that is to say, at the rate of slightly under $\frac{1}{4}$ of a lb. per diem. We have the same difficulties with later weighings as in the preceding cases, and for similar reasons. Figures concerning the half-bred ewe lambs continue, however, comparable, as all were kept for breeding purposes.

TABLE IX.

Showing Growth and Development of 145 Lambs by South Down Rams out of Merino-Dorset-Horn and Merino-Shropshire Ewes.

Dates when Lambs Dropped	First Weighing. June 29.					Second Weighing. August 11.				Average Increase in Weight.		Third Weighing. October 9.					Average Increase in Weight.		Average Increase in Weight.		Fourth Weighing. October 25.					Average Increase in Weight.		Average Increase in Weight.	
	Numbers.	Length.	Height.	Girth.	Weight.	Length.	Height.	Girth.	Weight.	For 43 Days	Per Day.	Numbers.	Length.	Height.	Girth.	Weight.	For 50 Days	Per Day.	For 102 Days	Per Day.	Numbers.	Length.	Height.	Girth.	Weight.	For 75 Days	Per Day.	For 118 Days	Per Day.
		in.	in.	in.	lb.	in.	in.	in.	lb.	lb.	lb.		in.	in.	in.	lb.	lb.	lb.	lb.	lb.		in.	in.	in.	lb.	lb.	lb.	lb.	lb.
WETHERS—																													
April 23 to 30	11	20.18	19.73	25.92	42.09	23.45	23.18	27.86	60.64	18.55	0.43	6	26.33	25.50	34.33	90.50	29.42	0.50	48.34	0.47	3	26.33	26.50	33.50	81.30	29.66	0.40	44.49	0.38
May 1 to 15	51	19.63	18.87	24.05	36.62	23.05	22.67	27.08	55.48	18.86	0.44	27	26.07	25.60	34.85	95.15	34.54	0.59	54.56	0.53	22	26.86	26.34	33.50	84.45	34.70	0.46	52.00	0.44
May 16 to 31	12	18.25	17.50	22.75	31.66	22.42	22.17	26.21	51.13	19.47	0.45	4	26.00	24.50	33.00	92.13	36.50	0.62	58.25	0.57	6	26.42	25.75	32.58	79.42	33.92	0.45	51.50	0.44
After May 31	3	16.67	17.33	19.67	23.17	21.33	21.83	24.33	46.50	23.33	0.54										3	27.33	27.17	32.83	90.67	44.17	0.59	67.50	0.57
Total	77	19.38	18.72	23.86	36.10	22.94	22.63	26.95	55.19	19.09	0.44	37	26.40	25.46	34.20	94.05	33.42	0.57	53.95	0.53	34	26.78	26.32	33.34	83.79	34.96	0.47	52.62	0.45
EWES—																													
April 23 to 30	5	19.80	20.20	23.80	33.80	22.90	21.50	27.30	49.60	15.80	0.37	1	29.00	26.50	34.00	95.00	30.54	0.52	47.00	0.46	4	26.25	25.50	32.13	76.00	30.13	0.40	45.76	0.39
May 1 to 15	54	19.31	19.47	23.63	35.11	22.85	22.28	26.70	53.30	18.19	0.42	19	26.68	25.79	33.71	91.71	32.58	0.55	53.34	0.55	33	26.14	25.44	33.48	81.95	32.11	0.43	49.09	0.42
May 16 to 31	9	18.33	18.39	22.01	31.88	22.89	22.61	25.94	51.29	19.39	0.45	1	28.00	26.50	34.00	88.00	27.00	0.46	45.50	0.45	7	26.79	23.79	33.57	83.57	32.50	0.43	51.50	0.44
After May 31																													
Total	68	19.24	19.45	23.51	34.59	22.86	22.26	26.71	52.76	18.17	0.42	21	26.86	25.86	33.74	91.79	32.21	0.55	51.76	0.51	44	26.25	25.50	33.37	81.67	31.99	0.43	49.17	0.42
GENERAL AVERAGE OF WETHER AND EWE LAMBS. Total	145	19.31	19.06	23.69	35.39	22.90	22.46	26.81	54.07	18.68	0.43	58	26.55	25.60	34.03	93.23	33.23	0.56	53.13	0.52	78	26.18	25.86	33.36	82.58	33.28	0.44	50.67	0.43

NOTES ON TABLE IX.

This table shows the development of 145 lambs by South Down rams out of half-bred Dorset-Horn and half-bred Shropshire ewes, from June 29 to October 25. The increase in weight in the first period was not high, corresponding roughly to that of the half-bred Shropshire lambs out of Merino ewes. This is attributable to the fact that there were in this flock a large number of twin lambs, that tended at first to bring down the general average. Ultimately, however, we got some of our finest lambs from this flock. During the first period wether lambs increased at the rate of 9-20th of a lb. per diem, with a maximum of 21 lb. to 51 lb.—that is to say, at the rate of 7-10th of a lb. per diem ; and a minimum of 52 lb. to 62 lb.—that is to say, at the rate of not quite ¼ of a lb. per diem. During the same period ewe lambs increased at the rate of 2-5th of a lb. per diem, with a maximum of 32 lb. to 58 lb.—that is to say, at the rate of 3-10th of a lb. per diem ; and a minimum of 37 lb. to 43 lb., or barely 3-20th of a lb. per diem. With this flock, both for wether and ewe lambs, we have the same difficulty for later weighings, as both were sold from time to time as they became available.

GRAPHIC REPRESENTATION OF DEVELOPMENT OF LAMBS.

Finally, to put the results graphically, so that their general bearing may be grasped at a glance, I have traced the progress of the ewe lambs on squared paper. I should, of course, much like to have been able to include wether lambs as well. Unfortunately, the irregularity of the later weighings that has already been referred to on several occasions does not permit of this being done. Had it been possible to include the wethers, the results would, no doubt, have been more showy, but not otherwise materially altered, so far as the position of the different crosses is concerned.

In the graphic illustration herewith the vertical lines

indicate the weight of the lambs in pounds, whilst the horizontal lines indicate the period of time during the course of which the lambs were under observation. It will be noted that half-bred Dorset-Horn lambs head the list, with half-bred South Downs second, and half-bred Shropshires third. I have not included lambs from half-bred ewes, because, as has been seen, they are open to the same objections as wethers from Merino ewes. Finally, for purposes of comparison, I have shown the development of a few pure-bred Merino lambs that were kept under observation at the same time. The extreme regularity of the development of the Merino lambs is worth noting.

I do not pretend to think that these results are likely to be reproduced under all circumstances ; I merely give them as the results of a single season, and that, too, a season by no means too favorable to the development of early lambs. I propose continuing these tests of the different crosses during the present season, and I trust to be so situated as to be able to carry them out with greater regularity. It should not be forgotten that purely experimental tests of the kind need to be watched with much care, and that they absorb time and labor that we are, unfortunately, not always in a position to give. I recognise that it is our business to carry out experimental work ; and in the course of the last three years I have endeavoured to give effect to this, as much as our circumstances have permitted. It may be stated that our facilities in this direction are not as great as they might be. It may, perhaps, be objected that our results do not do justice to some of the breeds tested, on the grounds that our rams are perhaps not of the best. To this I would reply that we keep good flock rams, equal to any of the average sold for purposes of crossing. If some breeders are dissatisfied with our results, let them present us with rams of their own choosing, and we shall be quite prepared to test them fairly on our flocks.

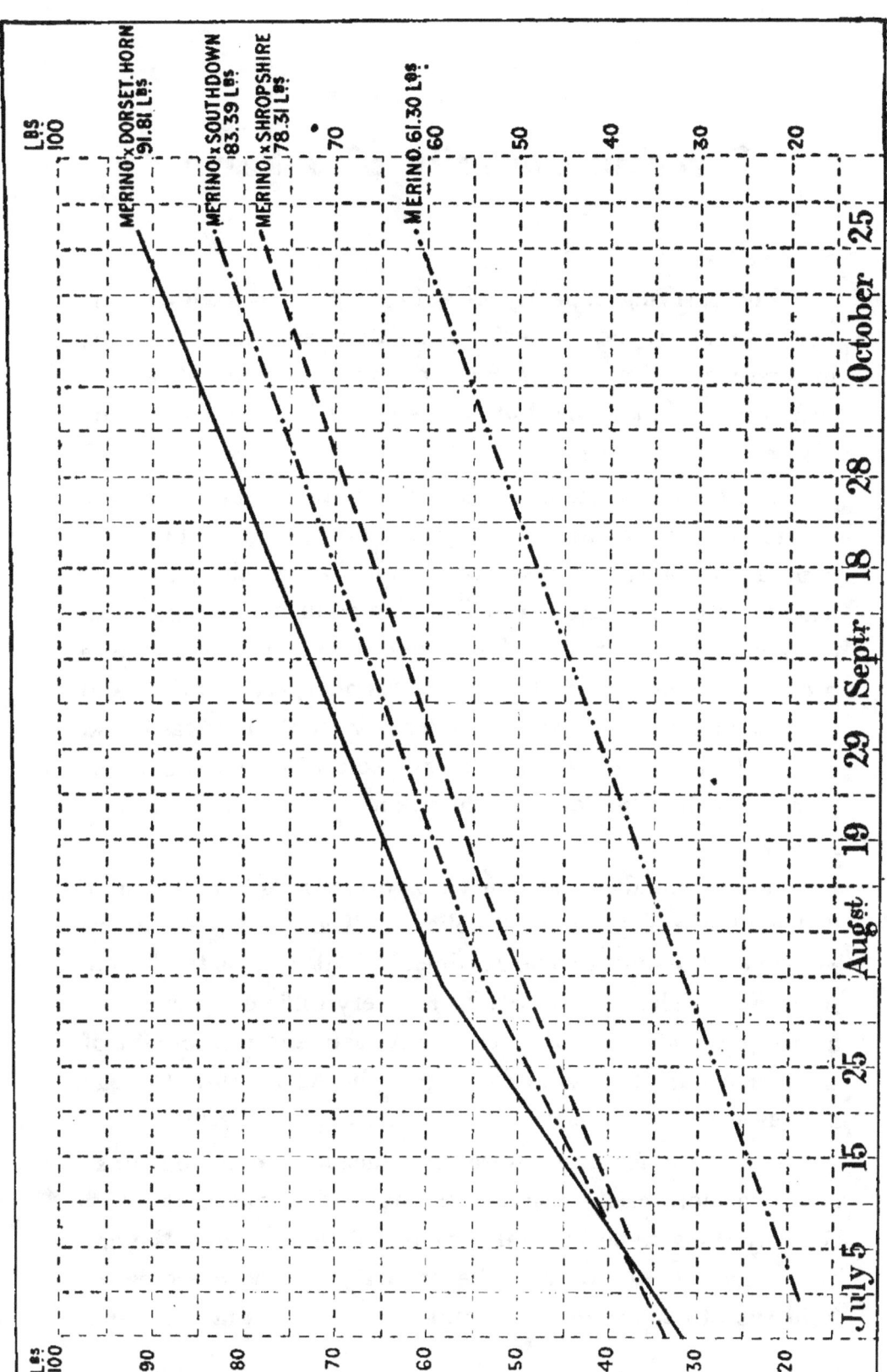

Graphic Representation of Development of Ewe Lambs in 1906.

Carrying Capacity of a Farm.

The carrying capacity of a farm is an exceedingly important matter, and one which it is safe to say is not generally understood. On some farms and stations more sheep are kept than ought to be, but it is equally true that on many places too few are kept.

Where overstocking is carried on, the results are in a general way very unsatisfactory, for not only are the sheep kept in anything but good condition, but at shearing time the result of the clip is usually disappointing. I know of cases where a number of sheep were kept, which afterwards proved to be more than the country should have carried, and when the number had been considerably reduced, it was found that not only were these in better condition, but they cut more wool than the larger number, and the clip realised more money.

It also stands to reason that to carry 200 sheep where 300 could be kept is bad policy. Therefore, the happy medium is the object to be attained. But what is the happy medium ? This has always been a very difficult question to answer, and, owing to the use of fertilisers and in the light of the recent experiments in feeding at the Agricultural College and elsewhere, the answer becomes more difficult still.

Before the fertilisers were introduced the carrying capacity was more easily defined, as the season was practically the only thing to be taken into consideration. Since the use of the fertilisers, however, the capacity has been increased from time to time until it is quite safe to say that in many cases three times the number of sheep are kept since they came into use, and on large areas of poor land, where previously it was impossible to grow wool, flocks are now seen.

The increase in the carrying capacity of the land as a result of the fertilisers is no doubt due to the stimulating effect on the natural growth during the fallow season, for not only is the stubble there to feed on, but the early autumn rains bring up a much more abundant herbage than used to be the case before fertilisers came into vogue.

One other important factor in the carrying capacity of the land, and one that must materially affect the export lamb trade, is the growing of catch crops. This practice is to some extent new to South Australia, but farmers are gradually realising the advantage of providing green feed for the ewe at lambing time. There can be no doubt that this is the right course to adopt, for, unless the ewe is provided with suitable food, she cannot be in a fit condition to properly rear a lamb.

Some of the best crops to grow are rape, barley, rye, mustard, vetches, oats, and even wheat may be taken advantage of in feeding the ewe. It is sometimes held against these crops that in a dry season they often fail, but when such fast growing crops as these fail natural herbage will also be wanting, but in any case there will be the satisfaction of having these crops some weeks earlier than the natural grasses.

Under the conditions which prevail North of Adelaide, and sometimes even in the south, the farmer must have other methods of tiding over the difficulty in a dry season when neither the fodder crops nor natural grasses have come on well.

At the Agricultural College and on other farms, farmers are making a good food for ewes by chaffing straw and mixing it with molasses. The labor involved in doing this is small, and the results obtained in a dry time will amply justify the practice. But even straw chaff and molasses are hardly sufficient food on which to raise early lambs. If the farmer wants to keep the ewe in such condition that the lambs will come away from the start, he will have to feed on something better, and should give the ewes one pound per day each of crushed barley, crushed oats, or half a pound of either of these with half a pound of bran, which quantity will be sufficient to keep the ewe in good condition. Roughly speaking, a pound of this food will cost about a penny, thus the ewe can be kept during the early part of the milking season at a cost of, say, 7d. per week. This kind of feeding will only be necessary when the conditions are such that lambs are suffering from want of green feed.

In times past, when lambs were being sold at 4s. and 5s. per head, this course of treatment was not justifiable, but at the present day, when they are selling at from 10s. to 15s. each, the farmer can afford to be more generous, and at the outside he will have to feed for not more than three or four weeks, which would only cost a couple of shillings per ewe, and this amount will be more than made up to him by the extra value of the lamb which has been carefully nourished from its birth.

Again, it must be clear that the ewe will be in a very much better condition at the end of a dry period through having been well fed whilst suckling the lamb, consequently the advantage to the ewe if for sale at the end of the lambing season is very great, and even if not for sale, the effect of its having been kept in good condition will be shown in the extra quantity of the wool produced.

So far, then, as the carrying capacity of the farm is concerned, the farmer must see that when growing early lambs, provision should be made for them in the shape of early feed, and if the season is a bad one, recourse must be had to some means of artificial feed, for, in the face of the prices ruling to-day, it will not pay to neglect newly dropped lambs.

Although the foregoing remarks have referred chiefly to ewes and lambs, they apply with equal force to wethers that are being fattened for market.

Until, however, farmers have had more experience in growing such crops as those mentioned it will be well for them to go slowly in the matter of materially increasing their stock.

Therefore, in the light of the foregoing the answer to the question, "What is the Happy Medium?" must be left to the farmer to work out for himself. If he goes slowly and feels his way the problem will gradually solve itself.

The Fat Lamb Raising Industry.

————

The breeding of fat lambs for market is no new occupation, for it has been carried on in a limited way for many years, though the results were none too satisfactory at first.

Although the business has been carried on for some considerable time in New Zealand, it was not until comparatively recently that the first consignment of lambs was shipped to the old country from South Australia. Since then the whole aspect of lamb raising has changed for the better, for it is now a matter of history that the industry has gone ahead by leaps and bounds, and this is one of the best proofs that can be given of its profitableness to the farmer. This year we shipped no less than 251,569 lambs from South Australia, and there is no saying to what extent this number will be increased as time goes on, because everything points to freezing as having become firmly and definitely established.

The market in Great Britain seems to be almost unlimited, but so far the continent has taken none of our lambs. There is every reason to believe that in the near future things will be different in this respect, and that Germany especially will absorb large quantities of Australian-grown meat, for there is no doubt that in that country the meat production is extremely short.

There is, therefore, little reason to think that for many years to come the fat lamb raising industry will be anything but very profitable to the breeder.

It is readily admitted that the whole thing is just in its infancy, and it appears that to a large extent the success or failure as a matter of business is in our hands, and it may be taken for granted that if we on our part breed the right class of lambs, the shippers on their side will see that none but those fitted for export will leave our shores.

There is much yet to be learnt, and, as the business so far has mostly been in the hands of the agriculturist, who, as has been pointed out, was in many cases comparatively inexperienced in the keeping of sheep, it will readily be granted that, as he becomes more experienced, the results will be better all round.

Then, again, we have to a large extent been in the dark as to what weight of lamb is best for freezing, for, while some years ago it was generally understood that a lamb of from 36 lb. to 40 lb. in weight was the ideal, the prices obtained for both lighter and heavier weights suggest that there has really been no standard weight which can be set up as the correct one.

So far very little attempt has been made at any sort of uniformity of either type or weight, but no doubt these matters will receive the attention they deserve as experience of the wants of consumers becomes more general.

The report of Mr. Pope (the Manager of the Government Produce Department) on the fat lamb raising industry, which was published in April, is so full of useful information that I asked and obtained that gentleman's permission to reprint it, and I am sure that its perusal will be both interesting and instructive to breeders.

Flock of Merino Sheep.

Sheep for the Farmer.

At the outset it should be stated that this chapter has more to do with the small farmer, and more especially with the ordinary wheat farmer who has had comparatively little experience with sheep, than with the larger farmer or squatter.

Naturally, in dealing with this subject, which covers wide issues, one dare not attempt to lay down any hard and fast lines, as there are so many different and ever varying conditions to be considered—different conditions as regards locality, quality of land, and whether fat lamb raising or some other branch of the business be the direct object; and varying conditions as regards altered methods of agriculture, such as the more abundant use of fertilisers, the growing of fodder crops, and the greater experience of the farmer generally, as well as in other directions, all of which make it impossible to dogmatise.

In order to make the matter as clear and simple as possible, I shall divide the subject under several headings, as well as districts, and as the greatest number of sheep are kept by farmers in the Lower North and on the Peninsula, and as fat lamb raising is, and looks likely to be, the most profitable branch of the industry, I shall first deal with it in relation to those districts.

Lamb Raising in the Lower North and Peninsula.

The carrying capacity of the farm is an essential matter. but, as it has already been dealt with, it will not be necessary to make any further reference to it here.

A matter of considerable importance is the kind of ewe to be kept for the purpose of lamb raising—a question that has given rise to a great deal of discussion.

Personally, I have for many years advocated the big-framed Merino ewes, but have always admitted that cross-bred ewes were unquestionably better mothers, and, seeing that it is impossible to get anything like a sufficient number of crossbred ewes, coupled with the fact that the Merino ewes are always obtainable, I still feel justified in adhering to my old opinion.

Although the wool of the ewe is not the most important matter, still it cannot be overlooked, and, while during the last few years crossbred wool has been selling at extremely high prices, and for that matter, Merino too has been selling above normal value, the experience of the past teaches that crossbred wool is much more affected by the ups and downs of the market than Merino, and when Merino wool drops in value, crossbred wool is more than likely to suffer to a greater extent.   In proof of this, many will remember that when, after the last boom, the wool market collapsed, ordinary Lower North Merino wool was selling at about 7d. to 8d. per lb., whilst crossbred was selling at 3½d. to 4½d. per lb., and, as the same thing has occurred on different occasions in the past, it is reasonable to suppose that history may repeat itself in this direction.   Again, there is the fact that, generally speaking, the Merino ewe will cut a heavier fleece than the crossbred, and in most localities will not require so much food.

The foregoing is based on the assumption that the farmer will buy his ewes rather than breed them, but which course is the more profitable is a matter on which opinions differ.

Professor Perkins, in his report points out. and rightly, too, that ewes bred on the farm are much more docile than station-bred sheep, which is a point greatly in favor of the farm-bred ewe, but it must be remembered that a large percentage of the farmers who have taken up fat lamb raising are comparatively inexperienced among sheep, and, as the breeding of ewes requires a certain amount of skill and experience, I think it will be some considerable time before the practice of breeding the ewe becomes general with the lamb raiser. Besides, by buying his ewes, the farmer gets a much quicker return than by breeding. For example, supposing a man has a farm, the carrying capacity of which is 100 ewes and lambs, the latter to be kept for, say, 4 or 5 months until they are ready for market. If he intends to sell all his drop, he can do so as soon as the lambs are ready, and thereby get his return in full. On the other hand, supposing he wants to breed his own ewes, it will be necessary for him to keep back, say 25 per cent. of his best ewe lambs which he will require to keep for another 12 months before he can put them with the rams ; and, thus, besides losing a portion of his more immediate return, it must be clear that if he had not reserved 25 per cent. of his lambs for breeding purposes *he could have kept more than 100 ewes to start with*.

However, possibly as farmers become more experienced with sheep, and have their farms subdivided into a greater number of paddocks, and in the light of the fact that crossbred ewes are better mothers than Merinos, and if bred on the farm are more docile than bought sheep. the practice of breeding his own ewes may become more general with the farmer.

Where crossbred ewes are bred, the question now is, which is the better cross to adopt. Here again we are faced

with a variety of opinions.   The Down or Dorset Horn cross,
or the Leicester or Lincoln, will answer the purpose ; but,
as there is not much advantage claimed for one class over
and above the other, I think it will become more a matter
of personal choice than anything else, although I am of
opinion that the two latter are the best.

RAMS.

The best breed of rams to be used for raising fat
lambs is a very important consideration, and, as early maturity
is one of the chief points to be aimed at, the Merino ram need
not enter into the question.   There are only about six British
breeds of rams in this State, viz., Shropshire Down, South
Down, Dorset Horn, Leicester, Lincoln, and Romney Marsh,
and of these I am inclined to favor the Downs (the Shrop.
from choice) or Dorset Horns.   Experience has proved
that these breeds have been well tested, and have been found
extremely valuable as getters of early lambs.   They are
justly famed for their early maturing properties, as well as
being prolific breeders.   Twins are not uncommon, and with
the Dorset Horn they occur more frequently than with the
Downs.   The South Down has many valuable attributes,
but it is not quite as large as the Shropshire.   The Leicester
comes here with a splendid reputation from New Zealand, as
well as from the old country, but as the conditions in the
countries referred to are so different from our northern conditions, there is no proof that this particular breed will give
better results than have been obtained from Dorset Horn or
Down breeds.   The Lincoln has also many valuable qualities,
but as a sire for fat lambs has not proved to be as good as the
others, whilst the Romney Marsh is essentially a wet country
sheep.

SELECTION OF PARENTS.

With cattle it is a recognised fact that the bull is half the herd and more, and with sheep it may be said that the ram occupies a position of equal importance, hence the type of ewe can be summed up briefly as a good, big-framed animal, growing a useful profitable fleece.

Whatever ram is used there is no doubt it will pay to use a pure-bred sire, although it may cost more money than a graded ram, for, whilst good results have been obtained by the use of graded rams, it is quite certain that the results would have been better had a pure-bred sire been used.

The price of a grade ram being generally less than a pure-bred one is a temptation to farmers to purchase the cheaper animal, but a moment's thought will show how little the difference in cost really amounts to. Say, for example, that a grade ram can be bought for two or three guineas, and the cost of a pure-bred be three or four guineas, the difference in the cost of production of each lamb is not more than 6d., and I am quite certain that the progeny of the pure-bred sire, when matured, will be worth at least one shilling more than those produced by the grade ram, everything else being equal, for it is a generally admitted fact that a pure sire gets a much more uniform stock than does a grade. In short, he transmits his good qualities to his progeny, whereas a grade sire is more likely to transmit his faults. And, after all, the difference mentioned in the cost of the production of the lamb only refers to one year, and, seeing that a ram may be reasonably expected to sire at least 150 lambs during his working life of three or four years, the difference would only mean about 1½d. per lamb extra. I am most emphatic on this point, and I am strengthened by the fact that breeders of every class of stock are at one in advocating a pure-bred sire.

It cannot be expected that the farmer can obtain for the ordinary flock ram price a perfect ram of any breed, but he can, and should, obtain one which is well-grown, bold, masculine-looking, robust, and vigorous, and which is rather a meat producer than a wool grower.

Where lambs are to be sold when they leave their mothers the amount or quality of the wool of the sire will not affect the selling value of the pelt (which. of course, affects the sale of the lamb) more than 3d. or 6d., whereas it is very easy to affect the meat value to the extent of a shilling or more, hence it is advisable in choosing a ram for fat lamb raising, to select one that will be likely to increase the weight of the meat rather than one that will produce the most valuable wool. This statement will have more force when it is remembered that rams carrying extra good fleeces, where everything else was equal, have not produced such good lambs as those carrying a short and indifferent wool. This is more particularly noticeable in the Downs and Dorset Horn breeds.

FARMERS SOUTH OF ADELAIDE.

In order to avoid repetition, I might say that a great deal of what has been written about the Lower North will apply to farms south of Adelaide, but the South-East must be dealt with separately. Whilst the main principles are the same as those obtaining in the North, the types used in the South will be somewhat different. The Leicester and Lincoln will be more common, as these breeds are suited to the wetter conditions which prevail in the South. The Leicester is unquestionably a quality sheep. It does not mature as early as the Downs or Dorset Horns, but cannot be said to be very backward in this direction. There seems little doubt that it will be used largely in producing half-bred ewes, which are extremely shapely and carry a good useful fleece. The good qualities of the Lincoln are well known. as it is an old-established breed in this State, and it, too, will doubtless be largely used in crossing with Merinos in the production of ewes for lamb-raising. There is no doubt that in these parts crossbred ewes will do better than Merinos. But even here the earlier-maturing breeds of rams will be kept for raising early lambs.

Referring to the South-East, in such districts as Mount Gambier and Millicent, where the conditions for fodder growing are so very favorable, the Leicester and Lincoln, as well as the Romney, will become favorites amongst the smaller sheep farmers.

In the more marshy districts round about Kingston, the Romney Marsh must come to the front, not only on account of its wonderful ability to withstand the diseases common to that country such as fluke, foot-rot, &c., but also on account of its meat and wool properties.

Sheep on the Farm Outside Goyder's Line of Rainfall.

Farmers in districts outside of Goyder's line of rainfall can hardly be depended upon to raise fat lambs for market on account of the frequent droughts with which they have to contend. Under these conditions, I strongly advocate the keeping of pure-bred Merinos, because during times of drought, when the lambs cannot possibly be fattened, and have to be sold, they would be equally as saleable as crossbreds, and if they have to be kept on the farm to replace some of the older sheep, there is no question as to the desirability of keeping Merinos.

Occasionally on account of an exceptionally good season, lambs are fattened in the districts referred to, and when a good Merino ram has been used, the results have been very satisfactory, and very excellent prices have been realised for them.

Of course, it has to be admitted that if Down or Dorset Horn rams are used, lambs in good seasons will be better still, but as good seasons are the exception and not the rule, I have no hesitation in adhering to my recommendation to keep Merino ewes and rams, as this breed is admitted by everyone to be the best for the drier areas.

The type of Merino most suitable for the district in question is undoubtedly the big-framed, bold, robust sheep, growing a lengthy staple, with good character, a fair amount of yolk, and with reasonable density.

Wool Carting on the Back Blocks.

The Large Pastoralist.

As has been already stated, this book is written generally for the small farmer. but, as much of the matter contained herein has reference to the larger pastoralist as well, I take this opportunity of adding a few further remarks. which will be based on what has already been written.

It must be quite clear that I am of opinion that for some years to come most farmers will not breed, but will rather buy their ewes, and as these ewes must come from the station, it stands to reason that those which will have the most ready sale will be the bigger framed ones. Thus, the pastoralists north of Adelaide, where the conditions are such that they can grow a big framed Merino carrying a valuable fleece, should undoubtedly breed this type. But independent of the fact that the ewe will be sought after by farmers, I am firmly convinced that this type of sheep will prove to be the most profitable all round one to breed. To those pastoralists in the hills districts where they cannot grow the larger framed ewes, the quality of wool, i.e., the fineness of fibre, is of more importance than to the northern squatter ; but even here I am satisfied that a good-sized frame is of great consequence to the pastoralists in the South as well as the South-East. In such districts as Bordertown, Naracoorte, and the like, where an exceedingly high quality wool is grown, I would only suggest that it will be in their own interests to give careful attention to size of carcase and length of staple, and if they look after the character of the wool, there will be very little danger of getting it too strong, and to those engaged in the industry in the wetter parts of the South-East, where crossbreds are kept, it will be well to remember that there are such long-wool breeds in South Australia as the Leicester and Romney Marsh, as well as the good old Lincoln.

Carting College Wool to Market.

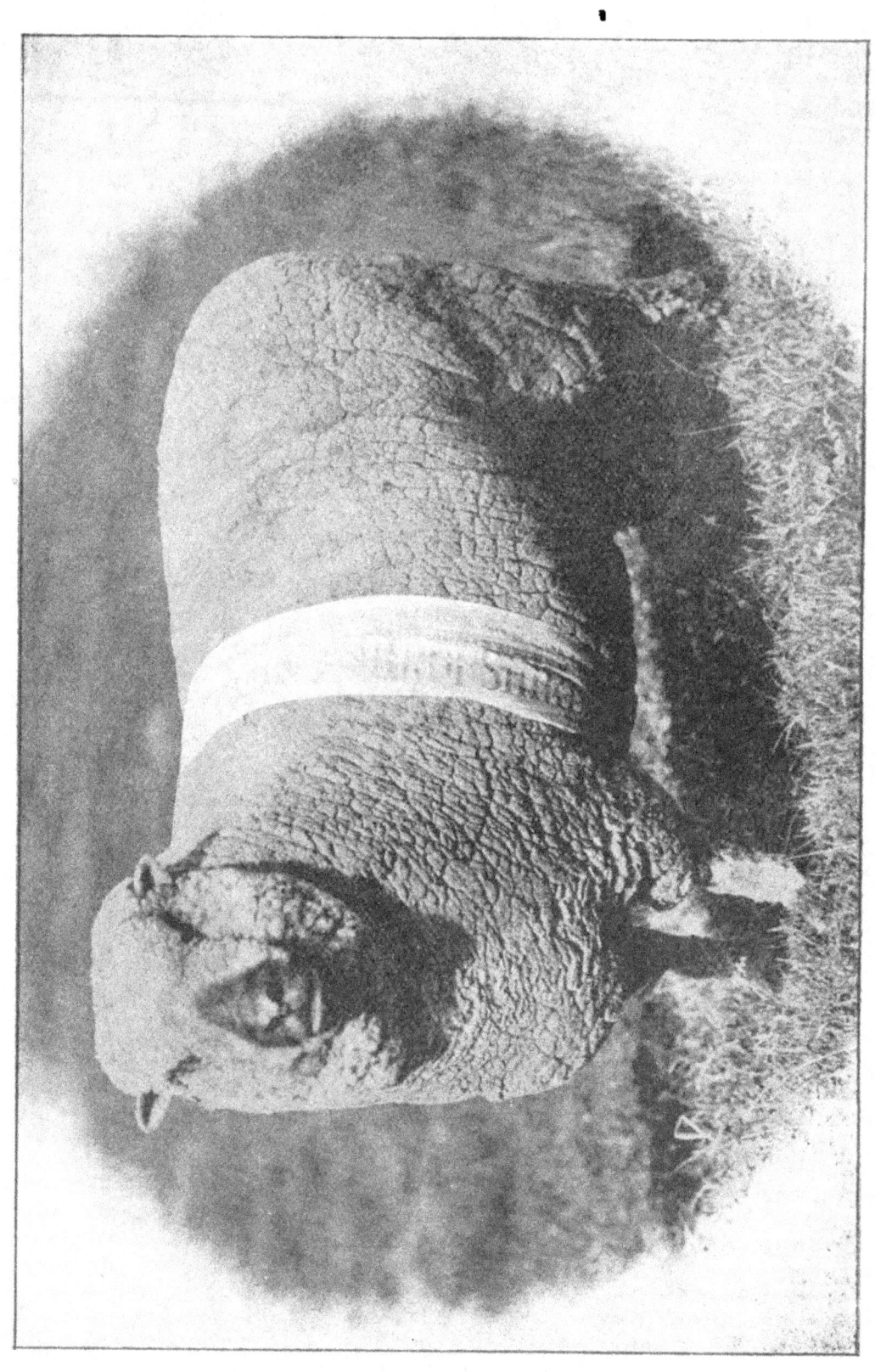

Shropshire Ram.

The Lincoln Ram.

This ram is very well-known. It has a good massive head without horns, is broad between the ears, with an arched poll. The ears are thick, and generally covered with a soft white hair. The face is intelligent, not too long, but broad and bold, and the skin is somewhat dark under the white covering of the hair. The lips are thick and the jaw deep and straight. The head is fairly well covered, with a topknot of good wool. The neck is long and thick at the base, tapering towards the head. The back straight and broad, although in some cases a little hollow. The body is long and the ribs well sprung. The shoulders set well forward, and the chest deep, while the sides are inclined to be flat. The legs are strong, thick, and wide apart. The rump is not very large, while the pelt is thick and coarse. It is one of the largest of British breeds.

Wool.—The wool, like the Leicester, belongs to what is known as the lustre class, and it is perhaps the longest and coarsest of its kind; the staples are thick and heavy and of great length, the crimps or undulations being wide apart. Although it is very coarse, it handles soft, and is of a very silky nature. It is made into a variety of fabrics, including serges, braid, furniture cloths, &c. The Lincoln-Merino cross produces a wool of exceedingly good weight, as well as good quality and value.

South Down Yearling Rams.

The South Down Ram.

This is one of the oldest of British breeds, and perhaps one of the most shapely sheep known. Like the Shropshire, it has a dark (sometimes called a grey) face and legs, although the hue is not so dark as the Shropshire. The head is smaller, very beautifully arched, and covered with a nice forelock, the forehead not very wide. The eyes are full and very bright. The neck thin at the head, getting larger towards the shoulders, although not long. The chest extremely well let down between the forelegs, and has a very massive, bold appearance. The shoulders are large, and on a level with the back, and the ribs are remarkably well sprung. The back is flat. The loin exceptionally broad and flat. The thighs very big. Legs comparatively short, while the pelt is fine and thin, but very pliable and strong It is a smaller sheep than the Shropshire.

WOOL.—The wool is finer and shorter than the Shropshire, and also lacks character. It has a mushy tip, and is of a harsh nature, and, like the shorter wool of the Shropshire, is mostly used for making flannel and hosiery goods.

Dorset Horn Ram.

The Dorset Horn Ram.

The Dorset Horn is also one of the old acknowledged pure breeds of British sheep. The head is large, with a pair of massive horns, after the style of a big Merino ram, but not so deeply corrugated ; the face is of good length, flat jaw, with a rounded broad muzzle, covered with a soft white hair ; the neck is short and thick ; the back long and straight ; terminating with a well-developed round rump, but the thigh not very deep ; the ribs are fairly well sprung and well rounded, the chest very deep and prominent ; the shoulders exceedingly broad, with a good fore-arm and brisket ; the legs short, thick, and strong, but, taken altogether, it is not such an attractive sheep to look at as some of the Down breeds.

WOOL.—The wool is somewhat longer than the South Down, but if anything a shade plainer, and distinctly lacking in character. It has, however, the advantage of being entirely free from black or grey hairs. The longer wool is generally combed, whilst the shorter can be used in the manufacture of flannel and hosiery goods.

Leicester Ram.

The Leicester Ram.

The head is comparatively small, and without horns, well set and clean, covered with short white hairs ; it is broad between the eyes, which are generally very prominent. The face is longer and sharper than the Lincoln ; the mouth wide, but the lips are not too thick ; the neck is thick towards the base and not very short ; the back is generally as level as a board, and the ribs rise with a sort of a rounded arch, which gives the body a regular barrel-like appearance ; the shoulders are broad and full and carried well forward ; the breast very full and deep ; the legs straight, and the bone thick, especially at the knee ;, the quarters are generally of good length and well filled out, the pelt thin and soft.

WOOL.—The wool differs from the Lincoln in that the staples or locks are not so broad, and being finer has more waves or crimps—the lustre is very apparent ; in fact, perhaps the most lustrous wool that is grown. The fleece of the cross between the Leicester and Merino is generally of a good weight and very excellent quality.

Lincoln Ram.

The Lincoln Ram.

This ram is very well-known. It has a good massive head without horns, is broad between the ears, with an arched poll. The ears are thick, and generally covered with a soft white hair. The face is intelligent, not too long, but broad and bold, and the skin is somewhat dark under the white covering of the hair. The lips are thick and the jaw deep and straight. The head is fairly well covered, with a topknot of good wool. The neck is long and thick at the base, tapering towards the head. The back straight and broad, although in some cases a little hollow. The body is long and the ribs well sprung. The shoulders set well forward, and the chest deep, while the sides are inclined to be flat. The legs are strong, thick, and wide apart. The rump is not very large, while the pelt is thick and coarse. It is one of the largest of British breeds.

Wool.—The wool, like the Leicester, belongs to what is known as the lustre class, and it is perhaps the longest and coarsest of its kind; the staples are thick and heavy and of great length, the crimps or undulations being wide apart. Although it is very coarse, it handles soft, and is of a very silky nature. It is made into a variety of fabrics, including serges, braid, furniture cloths, &c. The Lincoln-Merino cross produces a wool of exceedingly good weight, as well as good quality and value.

Romney Marsh Ram.

The Romney Marsh Ram.

This breed, although not such an attractive-looking sheep as the Leicester, is also bare in the face and without horns. The head is heavy and massive. The ears well set down on the sides, wide apart, and very thick. A topknot is the general covering for the forehead, which is wide. The eyes are inclined to be prominent, the nostrils thick, and the whole face, with its deep jaw, indicating strength. The neck is thick and fairly long, the back also long, but exceptionally wide on the loins. The thighs are particularly full, and the tail thick and heavy. The fore-quarters are inclined to be narrow, although the legs are exceptionally thick and the feet large.

WOOL.—The wool is of what is known as the demi-lustre class ; it is not so long in the staple as the Lincoln or Leicester, but shows a lot of character. It is also finer in fibre than the Lincoln, and when crossed with the Merino gives perhaps the most beautiful crossbred wool that is produced. The fleece, however, is generally lacking in weight.

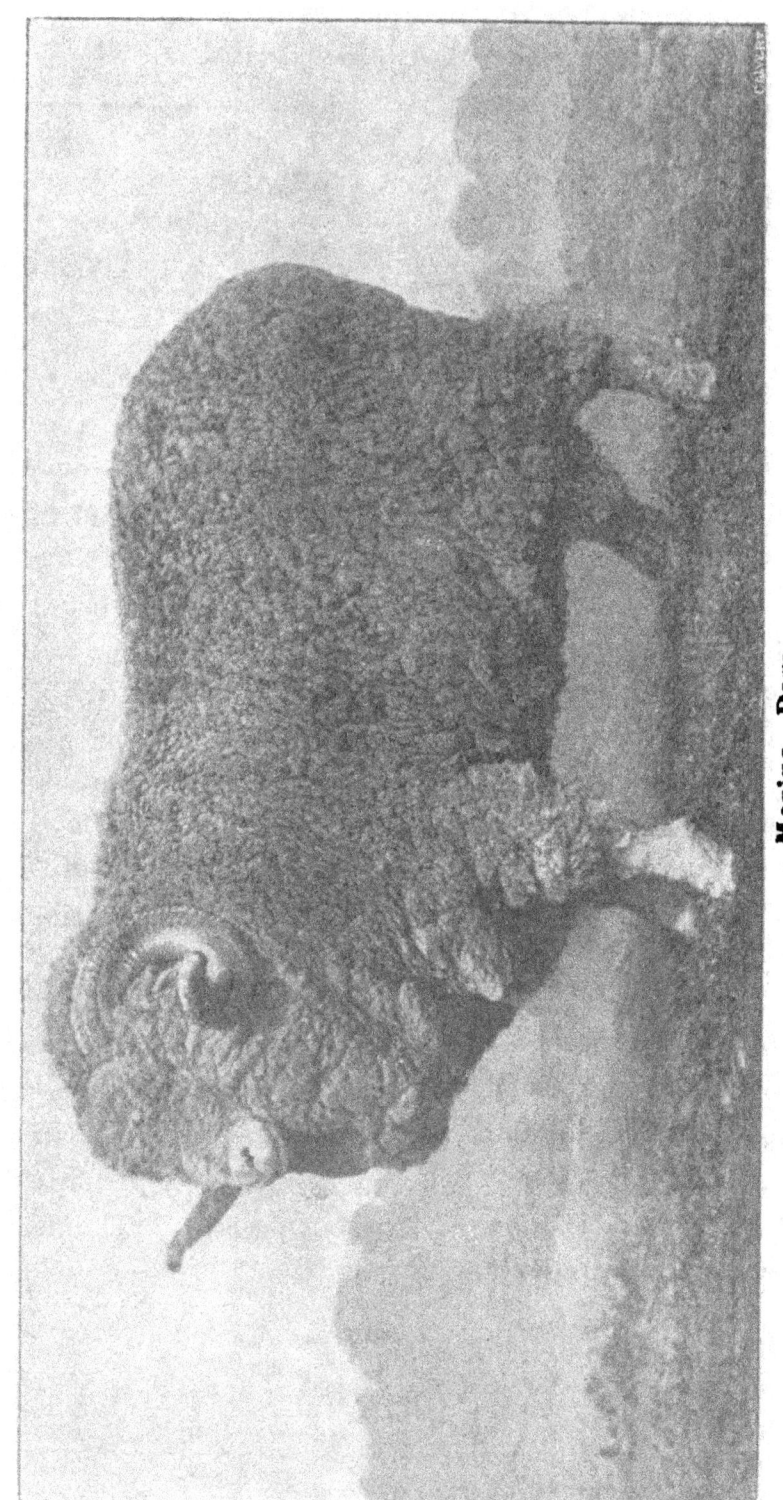

Merino Ram,

The Merino Ram.

Although the merino ram is not one which is largely used in the production of fat lambs for market, its influence in the breeding of ewes kept for the purpose is such that no book would be complete without a description of that ram, the main characteristics of which are as follows :—

The head not too big, but in proportion to the body, the poll slightly arched with horns well set back, taking a regular curve showing distinct and regular corrugations, and should not be too close at their base or to the cheeks, with sufficient room between them to allow for a growth of wool which continues well over the whole forehead, and in many cases on the cheek as well.   The ears of medium size, fine and soft. The eyes bright, intelligent, but mild and not too prominent ; the face of medium length and a trifle rounded, with two or three arches or wrinkles running from either side of the mouth.   The muzzle well rounded and nostrils wide.   The lower part of the face covered with a short soft velvety hair, and free from black spots.   The neck short and evenly set ; back straight and flat, ribs well sprung ; shoulders broad and full, well developed, with a thick fleshy fore-arm, broad, full, and deep ; brisket prominent and well let down between the front legs.   The whole front appearing massive, and having two or three folds, the bottom one the largest, the whole thing giving the appearance of a large apron.   The belly straight, not drooping, and the rump well rounded, quarters fairly long and full, the legs stout, with good bone and wide apart.   Knees flat, and the whole leg well covered with wool. Pelt thin, fine, soft, strong, and pink in color.

WOOL.—The wool of the merino varies on account of the type which is kept, as well as owing to the locality where they are depastured.   Thus some grow a short fine wool, others of medium length and quality, and some longer and stronger, but whether it be fine or strong, long or short, it should possess a distinct character, and have substance and body.   The merino wool possesses more yolk than the English breeds, and where the construction of the fibre is perfect it is found that the yolk rises to the top of the fibre and acts as a protection to the wool.

Merino Ewe.

Determining a Sheep's Age by the Teeth.

As a rule the teeth of a sheep are accepted as a guide to its age, but this rule is not infallible, which has been proved time and again. In fact, in many show rings the teeth are not accepted as a gauge by which the age of a sheep is determined.

The sheep, like the ox, has 32 teeth—eight in the lower front jaw (which are known as incisors), and 24 in the back jaws, which are known as molars.

The back teeth are placed both in the upper and lower jaw—six at the top and six in the bottom on each side of the jaw. There are no teeth in the upper front jaw, which is simply a pad.

When a lamb is born it may have two or four milk incisors, although these may not appear for the first four or five days. About three weeks later four more appear, making the complete number. During this period the temporary molars also make their appearance. The whole of the milk incisors remain for about 12 to 15 months.

No definite statement can be made in regard to the exact age when the permanent teeth first appear ; in fact, there is often a difference of three months between the times of some sheep getting these teeth on account of their individual development, but in a general way the following will be found to be substantially correct.

At about 12 to 15 months the first. two permanent incisors take the place of the two centre milk incisors. From nine to 12 months later, two more permanent incisors, one each side of the two already up, have replaced another two of the milk incisors, and from nine to 12 months later two more have done the same thing, making six permanent incisors, and 12 months later still, the two end ones make their appearance, and complete the full mouth of eight fully developed teeth. At five years and after, these teeth increase in length, and in some districts commence to break, although many of our Northern and Darling flocks do not show signs of breaking until two or three years later.

These permanent incisors are the teeth which are the generally accepted guide in determining the age. While the process referred to has been going on, the permanent molars have been displacing the temporary ones.

A sheep at the age of 12 to 15 months is known as a hogget or 2-tooth, at about two years a 4-tooth, after three years a 6-tooth, about four years an 8-tooth, or full-mouthed. After a sheep has developed a full mouth, only considerable experience will admit of its age being told, and in many cases it is not possible to say what the age really is. When the teeth begin to break or fall out, the sheep are termed broken-mouthed, and their general usefulness is rapidly on the decline.

Breeding.

Age When Rams and Ewes Should be Used.

Not many years ago it was common practice not to breed with either ewes or rams until they were four-tooth or over two years old. Of late, however, the practice has been altered, and it is not uncommon to use both ewes and rams as two-tooths, and the method on the whole has been fairly satisfactory, excepting in cold, wet districts where breeding with young ewes has frequently led to trouble.

To keep ewes for another year as maidens when they might be breeding lambs, and with no other return but the wool, cannot be called good business, especially if the conditions are such that the two tooth ewes can be used.

Whilst admitting that these young ewes are not such good mothers as older ones, it has been found that where proper care has been exercised, the results have quite justified their use. By proper care is meant that they must have plenty of feed and water, and that during lambing they are not neglected. On a small farm it is not at all difficult to go round them quietly, and where the ewe is down or in difficulty, to give the assistance required. If this is done, although the percentage may not be quite so large as from older ewes, still it will be found to amply justify the use of the young ewes in districts where it is not too cold and wet.

Experience also goes to show that where the conditions are favorable two-tooth rams can be used without any harm being done.   In fat lamb raising it is of a distinct advantage to breed with young rams, as their progeny mature much earlier than the progeny of the older rams.

Some time ago it had become a common practice to use ram lambs for this purpose when they were about six or seven months old, but where this was done a larger percentage of rams was necessary, as a ram lamb could not possibly serve so many ewes as an older one.

However, whether it is advisable or not to use ram lambs is a matter of opinion, but it is noticeable that many who adopted this practice have since discontinued it, and they are content to wait until the ram has further developed. There is no doubt that, where early maturing is an important factor, breeding with young rams is the right course to pursue.

Suitable Condition of Rams and Ewes for Mating.

Trouble has often been experienced on account of the ewes not taking the rams when mated, and the cause has generally been found to be that either the ewes or rams, or perhaps both, were too fat.   Ewes in good condition are more likely to take the ram than when they are very fat, and it is also true that rams are generally too lazy to work when they are in extra good condition.   It is, therefore, wise to keep them both from getting too fat previous to their being put together, as if the rams and ewes are in just good condition they will generally work all right.   If they are not working after being together for a day or two, it is often advisable to yard them for a few nights, when the results will be generally satisfactory.

PUTTING THE RAMS WITH THE EWES.

While most farmers know that five months, or to be more accurate, twenty-one weeks and three days, is the period of gestation with the sheep, many mistakes are made in this connection. It is often the case that no notice is taken of the date when the rams go with the ewes, with the result that the rams are left for an indefinite period, and, what is more serious, the lambs begin to drop before they are expected, and the result of this must be evident. About eight or nine weeks is a reasonable time in which to allow the rams to remain with the ewes. Although during the first five or six weeks most of the ewes will get in lamb, still it is wise to let the rams remain for about another month, as some ewes will no doubt be "returning," and will probably become pregnant. Assuredly it is better to have a lamb which is two or three weeks later than the others than no lamb at all, seeing that these later lambs will come at a time when there is sure to be an abundance of feed, and thus make headway very quickly.

There is a limit, however, to this, and it is a mistake to have lambs coming on at a time when the flush of green feed is past, hence about eight or nine weeks is a reasonable time to allow the rams and ewes to remain together. This applies more particularly to fat lamb raising, where a good start for the lamb is of such great importance.

The number of rams should be at least 2%, or two to the hundred ; some use a greater percentage than this, getting as high as three per cent., and no fault can be found with this method, for it is better to have too many than too few rams running with the ewes.

A very simple plan is to enter in a book the date on which the rams are put with the ewes, as well as when they are taken away, so that the farmer will have a fairly accurate idea when to expect his first as well as his last lamb.

The following breeders' table will no doubt prove useful in this connection :—

BREEDERS' TABLE.

TIME OF SERVICE.	MARES: 340 Days.		COWS: 283 Days.		EWES: 150 Days.		SOWS: 112 Days.		BITCHES: 63 Days.	
January	1	December	6	October	10	May	30	April	22	March	4
"	8	"	13	"	17	June	6	"	29	"	11
"	15	"	20	"	24	"	13	May	6	"	18
"	22	"	27	"	31	"	20	"	13	"	25
"	29	January	3	November	7	"	27	"	20	April	1
February	5	"	10	"	14	July	4	"	27	"	8
"	12	"	17	"	21	"	11	June	3	"	15
"	19	"	24	"	28	"	18	"	10	"	22
"	26	"	31	December	5	"	25	"	17	"	29
March	5	February	7	"	12	August	1	"	24	May	6
"	12	"	14	"	19	"	8	July	1	"	13
"	19	"	21	"	26	"	15	"	8	"	20
"	26	"	28	January	2	"	22	"	15	"	27
April	2	March	7	"	9	"	29	"	22	June	3
"	9	"	14	"	16	September	5	"	29	"	10
"	16	"	21	"	23	"	12	August	5	"	17
"	23	"	28	"	30	"	19	"	12	"	24
"	30	April	4	February	6	"	26	"	19	July	1
May	7	"	11	"	13	October	3	"	26	"	8
"	14	"	18	"	20	"	10	September	2	"	15
"	21	"	25	"	27	"	17	"	9	"	22
"	28	May	2	March	6	"	24	"	16	"	29
June	4	"	9	"	13	"	31	"	23	August	5
"	11	"	16	"	20	November	7	"	30	"	12
"	18	"	23	"	27	"	14	October	7	"	19
"	25	"	30	April	3	"	21	"	14	"	26

Breeders' Table *(continued)*.

Time of Service.	Mares: 340 Days.		Cows: 283 Days.		Ewes: 150 Days.		Sows: 112 Days.		Bitches: 63 Days.	
July 2	June	6	"	10	"	28	"	21	September	2
" 9	"	13	"	17	December	5	"	28	"	9
" 16	"	20	"	24	"	12	November	4	"	16
" 23	"	27	May	1	"	19	"	11	"	23
" 30	July	4	"	8	"	26	"	18	"	30
August 6	"	11	"	15	January	2	"	25	October	7
" 13	"	18	"	22	"	9	December	2	"	14
" 20	"	25	"	29	"	16	"	9	"	21
" 27	August	1	June	5	"	23	"	16	"	28
September 3	"	8	"	12	"	30	"	23	November	4
" 10	"	15	"	19	February	6	"	30	"	11
" 17	"	22	"	26	"	13	January	6	"	18
" 24	"	29	July	3	"	20	"	13	"	25
October 1	September	5	"	10	"	27	"	20	December	2
" 8	"	12	"	17	March	6	"	27	"	9
" 15	"	19	"	24	"	13	February	3	"	16
" 22	"	26	"	31	"	20	"	10	"	23
" 29	October	3	August	7	"	27	"	17	"	30
November 5	"	10	"	14	April	3	"	24	January	6
" 12	"	17	"	21	"	10	March	3	"	13
" 19	"	24	"	28	"	17	"	10	"	20
" 26	"	31	September	4	"	24	"	17	"	27
December 3	November	7	"	11	May	1	"	24	February	3
" 10	"	14	"	18	"	8	"	31	"	10
" 17	"	21	"	25	"	15	April	7	"	17
" 24	"	28	October	2	"	22	"	14	"	24
" 31	December	5	"	9	"	29	"	21	March	3

A Group of Dorset Horn Ewes.

Care and Comfort of the Sheep and Lambs.

CLIPPING ROUND THE SHEEP'S EYES.

In grass country when the wool begins to get a fair length, it is a good thing to muster the sheep and shear the face all round the eyes, in order to prevent the grass seeds from getting into the eyes and causing blindness. This is so important that, if necessary, it should be done two or three times during the year. It stands to reason that if sheep are blind they cannot see what they are doing, and, therefore, do not get the feed they should, and, consequently, they do not put on the condition they otherwise would.

When the sheep are yarded for the purpose of shearing the head and face, it may be found that their toes or clouts have grown abnormally long. These, if not shortened, will cause lameness ; and therefore, it is a good thing to cut them back with a knife.

CLIPPING ROUND THE PIZZLE.

Occasionally it is found that the wethers get inflammation of the pizzle caused by the stoppage of the urine owing to the growth of wool on the belly. The cure for this is to clip the wool all away and wash with a weak solution of Soda.

Whilst cutting the wool from the belly it is extremely important not to cut the small hairs which protrude from the penis as they act as a conducter for the urine, and thus assist in keeping the wool clean which will grow again on the shorn part.

DAGGING.

Whilst the sheep are still in the yard it may be found that they are daggy, in which case the dags should be clipped off, as this will greatly add to the sheep's comfort, and save wool at shearing time.

CARE OF EWES WHILE LAMBING.

A sheep man coming from the old country to Australia has always been struck with the evident carelessness that is exercised with the ewes when lambing ; but whilst in many cases nothing short of cruelty was practised in this respect, the fact must not be lost sight of that where the paddocks have been almost unlimited in size, there was no chance of the ewe receiving that attention which the home farmer was in the habit of giving. Then, again, the price of sheep did not warrant the expense of keeping a special staff of men for this purpose ; thus, if a ewe lost her lamb, or for that matter lost her own life when lambing, nothing at all was thought of it. With smaller farmers this want of care has been very apparent, and, without discussing the ethical view of the question, I would point out that as conditions are now totally different from what they have been in the past, the losing of a lamb at the present time means a loss of, say, 10/ or 12/, whilst if the ewe also dies it means as much again and more.

It is no excuse to say that during lambing time the farmer is too busy to devote any time or care to the ewes. Taking for granted that the farmer is keeping his sheep for the purpose of making money, it will surely pay him to give a little time and attention to the care of his ewes at this period, as by doing so he will increase his percentage of lambs considerably, for there is no doubt that a lot of the poor percentages we hear about are at least partly due to the neglect of the ewes at this important season.

No matter whether the ewes be old or young, it is the correct thing to frequently walk quietly through the lambing paddock, and be at hand to give the assistance that is often required, more especially, as has been before pointed out, in the case of young ewes.   Care should, however, be taken not unduly to disturb the ewes at this time.

Shearing Lambs.

In the event of the lambs not maturing as rapidly as they should do on account of an adverse season, it is of great importance to have them shorn so soon as there is evidence of grass seeds becoming troublesome.   The importance attached to this can hardly be over-estimated, because after being shorn, lambs invariably improve and make a fresh start, whereas if they are worried by grass seeds, they cannot possibly go ahead.   Then again, when shorn, they are much less liable to become blind through grass seeds.   Blindness thus caused is a very common occurrence, as will be better understood when it is stated that 75 per cent. of the deaths of lambs in transit to the Government freezing works were caused by the lambs being blind.   Further, the lamb's pelt is very much lessened in value by the presence of grass seeds.

Lambs Must be Fed.

Whatever kind of rams or ewes are used in producing fat lambs, the best results cannot be obtained unless the lambs are well fed from birth until they are matured and ready for market.   Thus the ewe must be in such a condition before lambing that it will have a plentiful supply of milk with which to suckle the young lamb.   This is a matter of such importance that I emphasise it with all possible force.

Disposal of the Lambs.

This is a very important matter, and in a general way

where the ewes have been bought, it has been found advantageous to get rid of the whole drop at about the same time, although it may be necessary to send them in two or more drafts.   Objections are raised against this method on the ground of the cull lambs which it is argued will pay better to keep a year or more than to sell at the reduced price which would have to be accepted, but against this it may be pointed out that the cull lambs referred to are indifferent animals which it will never pay to feed, and the better course where there are only a few is to kill and eat them on the place, while if there are a good many the local butcher will generally be found to be a buyer.

Blowflies.

The blow fly has now become a very serious pest to the sheep farmer, so much so that many ewes and lambs are lost each year from this cause. The question has frequently been asked what is the best thing to do to cope with the trouble referred to, and the best answer is "Prevention is better than cure."

I admit, however, that even where the greatest care is taken to prevent the ewes getting blown, there is nearly always a percentage affected, but this will be dealt with later.

To prevent lambing ewes being attacked by the blow fly, it is absolutely necessary to have them breeched and crutched some time before lambing, and if the following hints are acted upon, the trouble will be avoided to a very large extent. Hoggets should be treated in like manner.

Bring the ewes in, say, three weeks to a month before lambing, and while breeching, use great care not to worry the ewe more than is necessary. It is a common but bad practice to catch the ewe by the leg. This should not be done, but she should be picked up carefully in the arms and set down. When this is done, commence on the udder and take the wool clean away in a strip of, say, three or four inches wide, right up to the anus. then on both sides to the root of the tail, the idea being to take away all the wool that is likely to be moistened by dung or urine, and all wool which is likely to be so affected should be clipped off as close to the skin as possible.

Another important point is to have the end of the tail clipped, as, after lambing it will frequently be found that some of the discharge will adhere, and is sure to attract the flies.

When the sheep are breeched and crutched, it is necessary to use some arsenical dip, and most satisfactory results have been obtained by the following method of using the dip :—

The solution was made strong and applied with a spray pump, with a quarter inch nozzle, as the dip was mixed so thickly that it would have choked the perforated jet of an ordinary spray pump.   This, besides being a very effective method, is a quick one, as a few plunges of the pump compresses sufficient air to spray several sheep.   A spray was made by using the finger on the nozzle.   One man held the ewe whilst the other took good care not to miss any portion likely to be affected.   The dip was forced well down on to the skin, as this is essential for effective prevention.

A few days after spraying the ewes they were examined, and it was found that, although the fly had again blown a few of the sheep, the eggs had died on the wool almost as soon as deposited.   On further examination later on, the results were found to be equally satisfactory.

Where sheep have already been attacked, it is absolutely necessary to clip the wool from all round and apply the dip.

In some districts the wethers and even the rams get fly blown, in which case it will be necessary to use the remedy already suggested.

Watering Sheep.

Frequent reference has been made to the necessity of the sheep being well fed, but it is not less important that they should be supplied with plenty of good water, and, as the purity of the water is of great consequence, I am strongly in favor of watering at troughs in preference to dams.

Stock Inspector Williams made exhaustive researches while he was in the South-East in this connection, and he proved conclusively that much of the disease that was prevalent in the district referred to was due to the stock polluting the water they had to drink, and this conclusion is quite feasible, when we think of the impurity that must naturally be caused by the droppings of stock while they are drinking in the dams.

If the Inspector's conclusions hold good in the South-East they apply just as forcibly in the North (although the conditions there are not such that the stock will so readily contract diseases as in the South-East). Then, besides the pollution of the water, there is the distinct waste (which is a very important consideration in dry seasons) through the sheep wading into the dams and the wool becoming saturated with water, which is carried away. Again, there is the loss which is frequently caused by the wool on the points, belly, and often on the sides of the sheep, being more or less injured by the mud in the dam.

The only argument used against the troughs is on the

ground of economy, but even were the cost twice as much, I am so convinced of their necessity that I would still advocate their use, and, after all, the cost of erecting the necessary windmills and troughs alongside the dam (which would have to be fenced) is comparatively small, and the upkeep, where reasonable care is taken, will not amount to much.

It is, of course, understood that when a farm is subdivided into small fields. the water must necessarily be supplied almost exclusively in troughs.

There is no doubt that if sheep drink plenty of water they will keep in much better condition than if they do not. In the winter time, owing to the moisture in the feed, they are not so likely to go to the trough, thus the salt lick, which naturally causes them to be thirsty, encourages them to drink.

A Lick for Sheep.

It is very important, especially in the summer time, that sheep should have some stomach tonic, and, as it is frequently used in the form of a lick, it will be instructive to know what Veterinary Surgeon Desmond has to say on the question.

"Veterinary Surgeon Desmond does not agree with mixing salt, bonemeal, and sulphate of iron as a lick for cattle and sheep. The salt and bonemeal should be placed in separate troughs. Sweet bonemeal for cattle should be purchased, as there is a great danger of communicating fatal diseases to stock through the medium of bonedust unless it has been prepared for the purpose. Sulphate of iron (green copperas) is much disliked by stock. They will not partake of food or water if this valuable tonic is added in large quantities. It is best administered dissolved in water. Where the water supply is in troughs hang two bags containing sulphate of iron on each end of the trough. Put the bag in the trough until the water becomes red, then remove ; put the bags again in the water when it loses its red color. Where the water supply is a running stream sulphate of iron must be given in the food supply in the case of cattle and sheep by artificial feeding. Dissolve 1 lb. of sulphate of iron in 25 gallons of water, and add a pint of this solution to the feed morning and night for cattle, and for sheep about one-sixth of this quantity. The sulphate of iron is a valuable tonic, and can be kept up for any length of time. Its administration to sheep would help to keep them free from internal parasites, and it would also improve the quality of the wool."

But where a mixture is preferred as a lick, the following recipe has been well tried, and I can thoroughly recommend it :—

> 40 lb. of salt.
> 1 lb. of sulphate of iron.
> 2 lb. of bonemeal.
> ½ lb. of sulphur.
> ½ lb. of lime.

The use of common salt alone has also been found of great benefit, and is certainly better than no lick at all, especially in the winter time.

Lamb Tailing and Cutting.

About four weeks after they are dropped is the usual time for ear marking and tailing all lambs and castrating ram lambs. Tailing with a knife causes considerable loss of blood, which unquestionably has the effect to a greater or lesser extent of checking the growth of the lamb for the time being; therefore, any effective method which does away with the use of the knife in tailing must recommend itself to everyone, and more particularly to the grower of lambs for market, where the necessity for the lamb getting away as it were from the very commencement, without any thing to retard its development is so essential.

Searing instead of cutting has now been going on for some little time, and I know enough about it to recommend its adoption (in the case of the small farmer, at any rate), but the greatest care must be exercised to avoid scorching the hinder parts of the lamb.

A very useful searing table is shown on page 97, with barrier, which effectively protects the lamb from injury by the hot irons.

Science has taught the value of disinfectants, and where the knife and ear clipper are used, this teaching can be well applied by the simple method of disinfecting all tools used in some antiseptic solution. A 5 % solution of ordinary washing soda, used hot, will sufficiently disinfect knives or other tools used.

As a further safeguard against blood poisoning the knife used had better be non-closing, the blade so well fitted as to leave no openings in which impurities can collect.   The instruments should be dipped after every few operations.

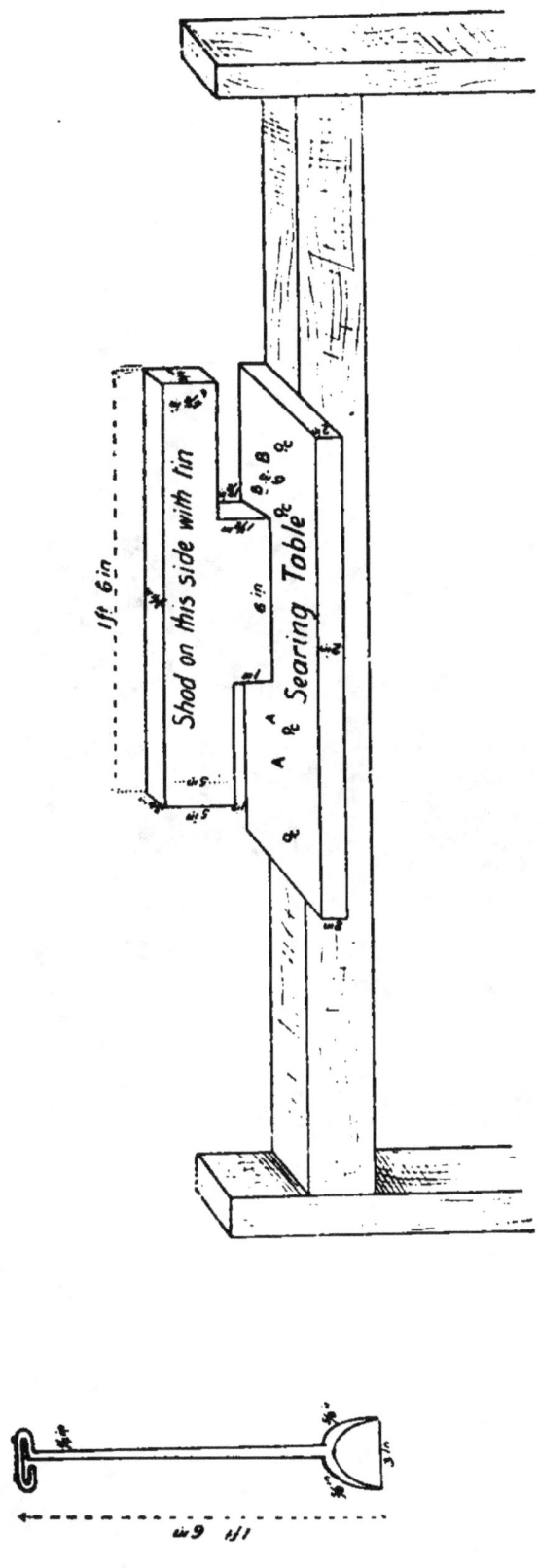

Searing Table.

NOTE. Right hand opening is 1¼ in., left hand opening is 1 in., for large or small tail.

Searing is done at B or A, lamb being offered at back of cross; operator, standing facing table, pulls the tail under either opening he desires.

Table should be of hard wood, and nailed or screwed to top of rail as shewn in drawing "C." Cross can be either hard or soft, but must be faced with tin or iron. The cross is made of ⅞ in. stuff, as it is thus a gauge for length of stump tail.

Searing Iron 1 ft. 6 in. over all. Handle for half way ¾ in., increasing to ⅞ in. at blade, which blade is also ⅞ in. in its thickest part.

The idea of this table came from Mr. Buttfield, at Pewsey Vale.

A Boundary Rider.

Selecting Fat Sheep.

The selection of fat sheep and lambs is no doubt a difficult matter, especially for the novice in sheep raising, and he must often experience disappointment with the result of the sales of his fats, so called. To the experienced grower or salesman, it is not difficult to class or draft fats for market, and perhaps the best way to assist the beginner will be to lay down the lines on which the experienced man goes to work.

First there is the general appearance of the sheep, which to a very large extent tells whether it is fat or otherwise. Then, after having taken a general look, the selector proceeds to handle the sheep over the loin and along the ribs. Following this he generally takes hold of the tail. Experience has taught him that if a sheep's loin is well covered it will carry a good kidney when killed—that is, a kidney that is well covered with fat, which usually affects the weight of the sheep to the extent of several pounds. It is very rarely that a sheep with what is known as a plain loin will kill well, although it may often have a good fat tail ; thus the tail is not an infallible guide. The sheep with a good loin will handle very solid on the back and across the hip bones. If the sheep's hip bones project in any way when handled, it cannot possibly be classed as prime. When the ribs of a fat sheep are as well covered as they ought to be, it will be difficult to feel each individual rib. A sheep which is not fat can also be detected by the parting of the wool right along the back. This want of condition gives rise to the stock salesman's term of " razor back."

With reference to lambs, they should be sold fat from their mothers if possible when they are termed in this State " spring lambs "; in many other places " suckers." Lambs coming in ready for market as late as January, having been weaned for some time and fed on lucerne, are classed on similar lines to fat sheep.

Erecting Yards.

In building sheep yards the following points will no doubt be helpful. To begin with, where possible, a suitable spot should be selected. A small rise in the ground is one great factor in keeping the yards dry in wet weather, as it allows drainage, and thus helps to avoid the ankle deep muddy yards that one comes across occasionally.

Then again, as sheep are more inclined to run up than down a slope, it is clear that it is advisable to build the race up the hill rather than down.

A race running westwards has its drawbacks. This is especially noticed when the work is done in the afternoon as the sun is then at the back of the drafter, and throws his shadow into the faces of the sheep which are coming towards him : this often causes the sheep to stop and jump back, whilst if the race runs eastwards the sheep, on seeing their shadows before them, will suddenly turn back and block the progress of those behind, thus the race should run north and south where practicable.

On page 102 is a sketch of very simple and effective drafting yards, together with an explanation as to the construction of same.

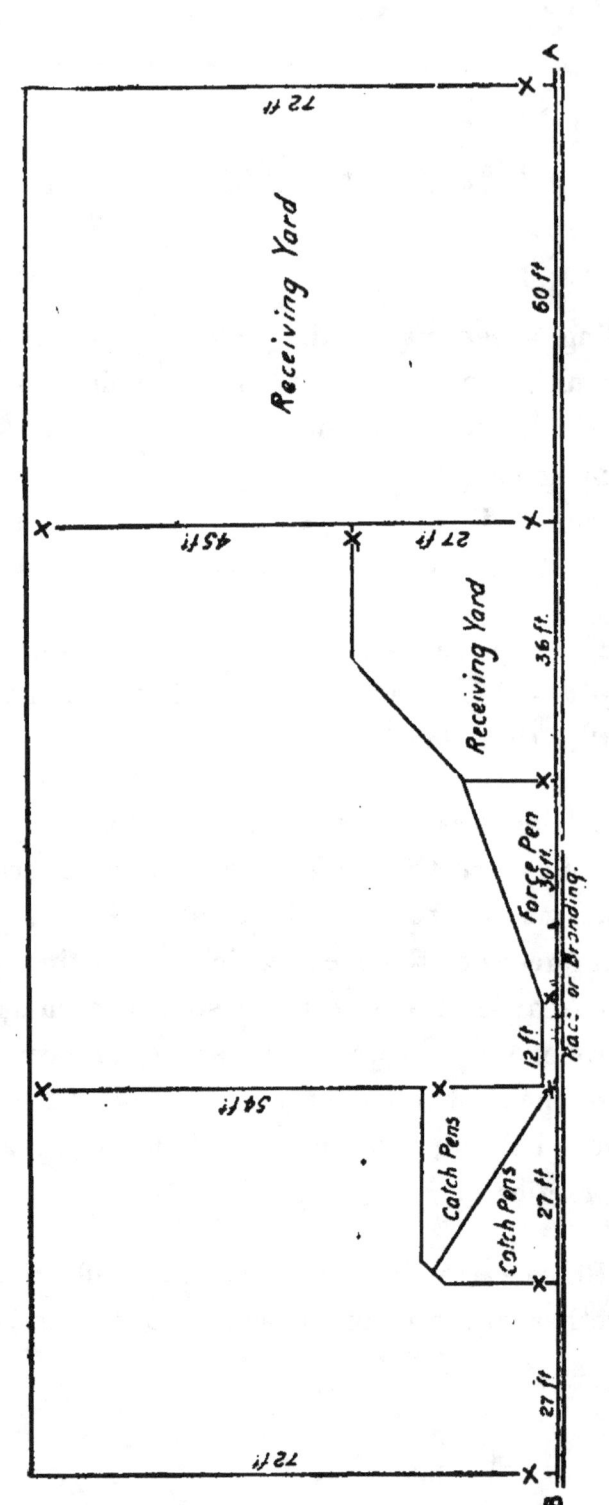

A to B represents an existing fence.   X represents Gates.

Scale—24 ft. 1 inch.

Handy Sheep Yard for Farmers.

Culling and Drafting.

Under the heading of "culling," all that need be said is that wherever it is practicable the inferior sheep should be taken out of the flock. This work, as regards wethers and dry ewes, is not difficult, but in the case of breeding ewes, great judgment is required, because through faithfully mothering a lamb or lambs the ewe reduces her own condition to such an extent that she might be culled out by an inexperienced farmer, although in reality she may have been one of the best of mothers. For this reason, shearing time is not the best for culling breeding ewes.

In selecting or drafting fats for market, it is not an uncommon thing whilst the drafting is going on to see men poking the sheep with sticks. This practice cannot be too strongly condemned, as every poke generally means a bruise, and most breeders know that in the export business any signs of disfigurement, such as bruises, are at once detected by the freezer, who then classes such carcases amongst those known as "rejects."

From the foregoing it will be seen that when sheep are handled in the objectionable way referred to, besides the breeder or exporter being the loser, the State in which they are produced must suffer in reputation, inasmuch as instead of getting full credit for producing sheep or lambs of the best description, a lesser credit has to be accepted because of the number of inferior grades.

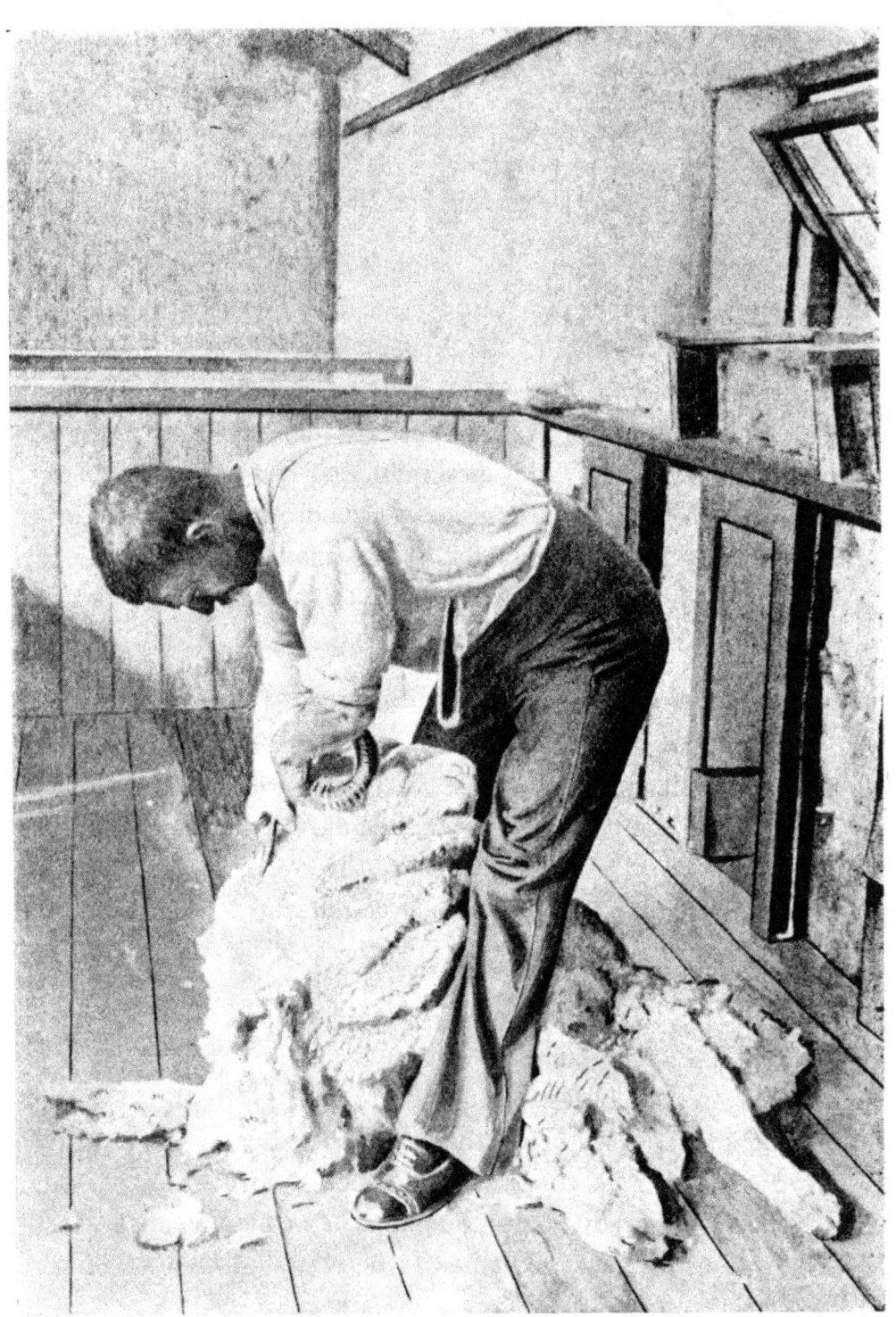

Hand Shearing.

Shearing.

The question to be faced by the farmer to-day in connection with shearing is not so much whether he will shear his sheep by hand or machine, but rather as to whether he will take his sheep to a central depot to be shorn, or shear them on his own place.

There is a movement on foot which has very much to recommend it, viz. : Contractors are renting shearing depots in various farming centres, so that the farmer who is handy can send along his sheep to be shorn. These depots are fitted up with shearing machinery, and a team of shearers is employed and kept so long as sheep are being brought into the shed. The advantages of this movement are very apparent, for it is quite possible for a farmer with his, say, 100 to 1,000 sheep to have the lot shorn in less than a day, whereas under the present system it frequently takes him a very long time, besides the inconvenience of carrying on the work at home.

The advantages will be more apparent when the ewes and lambs are being shorn, for it is very important to have these shorn as quickly as possible. In any case, whether the work is undertaken in the depots alluded to or on the farm by means of small motor shearing machines, or by the ordinary hand system, the following points should be borne in mind :—

When the shearer is selecting the sheep he intends to shear, it is most important that it should be lifted up from the sweating pen and not dragged out by the leg. This refers to all sheep, but more particularly to ewes, especially breeding ewes. There is more damage done by the sheep being dragged out by the leg than most people imagine, and the practice should be discountenanced in every possible way.

The shearer should take as much wool as possible from the sheep without cutting the skin or breaking up the fleece unnecessarily, and it is very important that as few second cuts be made as possible.

By second cuts is meant the wool that has been missed by the shearer in the first cut and taken off by a second blow of the shears. This wool either adheres to the fleece, which makes it unsightly and of less value, or falls beneath the spokes of the rolling table amongst what is known as the locks, and this is the lowest priced wool in the clip. As the wool of the second cuts is frequently what has been left from the best portion of the fleece, the reasons for avoiding them must be obvious.

By shearing the sheep clean, the grower gets to the full extent the benefit of the year's growth of wool.

To cut the skin more than is avoidable, besides being brutal, is bad policy, because when big patches of skin have been cut off, the sheep will fret, and will, therefore, not put on condition to the same extent as it might have done, besides which the presence of skin on the wool may affect its sale.

The floor of the shearing shed should be kept as clean as possible of all dirt, chaff, straw, &c., &c.

The following table will be. found very useful in making up shearers tallies :—

SHEARING TALLY READY RECKONER.

Sheep	25/- per 100		22/6 per 100		20/- per 100		17/6 per 100	
	s.	d.	s	d.	s.	d.	s.	d.
1	0	3	0	2½	0	2,	0	2
2	0	6	0	5	0	5	0	4
3	0	9	0	8	0	7½	0	6
4	1	0	0	10½	0	9½	0	8
5	1	3	1	1½	1	0	0	10½
6	1	6	1	4	1	2½	1	0
7	1	9	1	7	1	5	1	2½
8	2	0	1	9½	1	7	1	4,
9	2	3	2	0	1	9½	1	7
10	2	6	2	3	2	0	1	9
15	3	9	3	4½	3	0	2 .	7½
20	5	0	4	6	4	0	3	6
25	6	3	5	7½	5	0	4	4½
50	12	6	11	3	10	0	8	9
75	18	9	16	10½	15	0	13	1½
100	25	0	22	6	20	0	17	6

In each case the half-penny nearest to the actual fraction is shown.

Machine Shearing.

Dipping.

The importance of dipping in a district where sheep are affected by tick is now so generally recognised that the carrying out of such work is enforced by the various Governments, but were it not enforced it would pay the farmer handsomely to dip.

The tick, besides discoloring the wool, thus rendering it of considerably less value, naturally irritates the sheep to such an extent, that it cannot possibly thrive as it ought to, hence it does not put on the condition it otherwise would, and further, it does not grow its normal amount of wool.

In many districts, even where tick is not noticeable, dipping has been carried on with evident success, as the effect has been to thoroughly wash the skin as well as to kill any parasite which was either on the skin of the sheep or on the tip of the wool, and where a good reliable arsenical dip is used it acts as a tonic to the skin.

In dipping, it is essential to keep the bath well stirred, and it is also very important that the sheep should be kept long enough in the liquid for the wool to become thoroughly saturated, thus, if the bath happens to be a short one, the sheep should not be allowed to get out too quickly, because simply passing through a short bath would not in many cases be sufficient to do the good that is required, and if this matter is attended to there will be fewer complaints about the ineffectiveness of dipping.

Dipping is generally done immediately after shearing in order to save the sheep from being handled more than can be avoided, and when the work is well carried out little exception can be taken to dipping at that time.

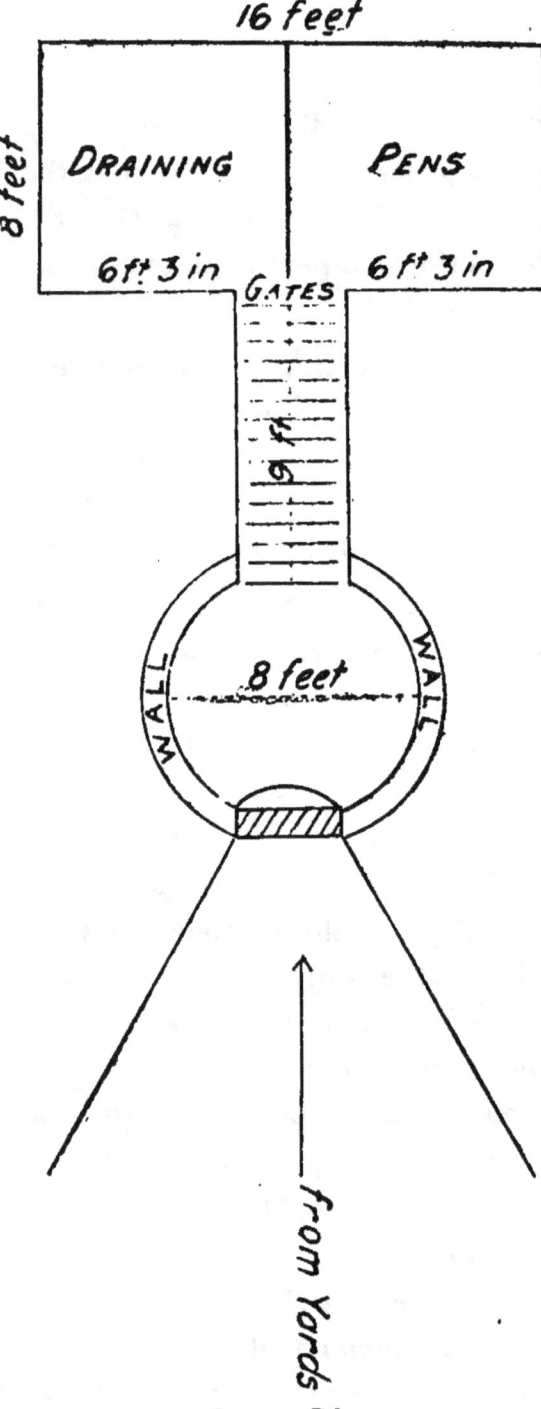

Sheep Dip.

In the draining pens, sheep stand on grating. The floor of yard under the grating may be of cement, with slope to tank to allow the dip to drain back.

Let one pen at a time fill with sheep. Then by the time you fill the other pen the first lot will be drained.

The slope from the tank to draining is 9 ft. long from bottom of tank and 3 ft. 6 in. wide with cross pieces for footholds.

The tank is 8 ft. wide and 7 ft. 3 in. deep inside. The wall should be built 1 ft. higher than the ground surface, to prevent the dirt tumbling into the tank.

The entrance to the tank should slope inwards at an angle of 45 degrees, having on the face a slab, say 3 ft. 6 in. square, for sheep to slip on.

Scale—8 feet to 1 inch.

Principles of Wool-Classing.

———

Although during the past few years much more attention has been paid to woolclassing than it had previously received, there are still those who have not a true conception of what the work really means, and in order that the principles may be thoroughly understood, this chapter which is largely taken from my previous book, is inserted.

In Australia the term " Woolclassing " is applied to the work carried on at the station at shearing time, when the wool is being prepared for market.

The object of classing is to get up the wool in such a way as to induce buyers to purchase at as high a price as they can possibly pay.

In order to obtain the best results, the classing must be done *skilfully, honestly, and carefully.*

If done skilfully, a greater amount of competition will be ensured ; if honestly, confidence on the part of the buyers will result ; and if carefully, the general appearance of the wool will be such as to attract attention.

SKILFUL CLASSING means dividing the clip into such lots as will best meet the requirements of the various sections of wool buyers (who, of course, represent the manufacturers), and in such a way as to enable them to most correctly ascertain its value.

In order that this may be clearly understood, it should be stated that all manufacturers do not require the same class of wool, nor, indeed, can they all make use of the same qualities.

For instance, "Combers" and "Carders" are two distinct sections of manufacturers, each using a different class of machinery, and their requirements vary according to circumstances, some manufacturers requiring "fine" or "medium" quality, others very "coarse" wool for either combing or carding. Some Combers only use a wool with a "sound staple," others are not so particular as to this quality. Some prefer a "broad stapled" wool, while others readily purchase wool which is "lean of staple."

Burry wool is such a source of trouble to some manufacturers that they will, if possible, avoid it, while others buy this class of wool quite freely. One particular section will only buy wool which is very light in condition, though, as a rule, buyers are not so particular in this respect, provided the wool is not too "heavy," and is classed in such a manner as to enable them to correctly estimate its "yield" when scoured.

The American section, who will only purchase a "light" conditioned wool, at present have to pay fivepence halfpenny per pound duty on greasy merino wool, sixpence on greasy long wool, and elevenpence per pound on all scoured wool imported by them into the United States.

Dealing more minutely with the varied requirements of buyers, we find that Combers want a "combing wool," i.e. wool of a certain length and strength; Carders use a "clothing wool," i.e., a short wool; manufacturers of light weight materials must have a wool of "fine" quality, i.e., wool, the fibres of which are of a very small diameter. Those who make heavier goods can use a "coarse wool," *i.e.*, wool of thicker fibre. Some Combers, more particularly the Bradford section, will only buy a "sound" wool, *i.e.*, wool, the staple of which is not readily broken during the process of combing, because, owing to the class of machinery they use, "tender" or "unsound" wool cannot be combed economically. This section also prefers a "broad-stapled" wool, *i.e.*, wool with the fibres grouped together in large clusters. Many Combers, especially those on the Continent, will readily buy a "tender" wool, even if it be "lean of staple," *i.e.*, stringy, and with its fibres in small clusters, because their machinery can comb that class of wool profitably. Some manufacturers, by reason of their particular requirements, can purchase burry wool with impunity, having an up-to-date plant for removing burrs, but in a general way burry wool is difficult to deal with, and many Combers will not touch it at all.

Although other examples might be given, the foregoing will be sufficient to show the importance of skilful classing.

AN HONEST " GET-UP " means classing the wool in such a manner as to inspire buyers with confidence, so that they may feel satisfied no attempt has been made to deceive by false packing, &c.

CAREFUL CLASSING means that the fleeces must be evenly skirted, and, should the necks or backs be taken out, they should be removed without tearing the fleece more than is absolutely necessary ; that the fleeces be neatly rolled and not tied with string, or even with part of the fleece unless it cannot be avoided ; that the pieces, bellies, locks, stains, &c.. have been properly looked after, and the pressing and other details of the " get up " have had due attention.

The foregoing applies more or less to woolclassing on all stations, whether Merino or crossbred, be they large or small, and even applies to small farmers, although, in their case, the sub-divisions will be comparatively few.

After due consideration has been given to these facts, the benefits of classing must be admitted.

Now that the importance of woolclassing has been dealt with, it remains to be shown how the grower is likely to get more money for his wool if properly classed.

In the first place, in all businesses it is generally admitted that competition is the life of trade, and when it is known that many of the classes of wool referred to are found on each station, the value of classing, in increasing competition amongst buyers, cannot be doubted, especially in the light of the fact that many large and important buyers often "pass" unclassed or badly got-up clips without making a valuation.

It sometimes happens that a buyer is compelled to purchase wool that he does not want on account of a "lot" being badly classed, or not classed at all, and yet consisting in part of that which he requires ; in consequence of this he will only give a reduced price, because he must allow for the possibility of loss in selling the wool he does not need, and also for the trouble and expense of repacking, &c.

Then, as classing assists buyers to correctly estimate the yield of a "lot" of wool, they are sure to stretch a point on that account, because they are making their valuation, as it were, with their eyes open, whereas, in valuing wool which has not been properly classed, they are at a considerable disadvantage. As will be seen in the chapter on wool-selling, a difference of one or two per cent. in the estimated yield means from one farthing to a half-penny per pound in the value of the wool, it is only reasonable to suppose that when there is the probability of an error of judgment occurring, the buyer will allow himself the benefit of the doubt, at the expense of the owner.

When it is also remembered that buyers have only a limited time in which to value the wool, this point will have all the more importance attached to it.

Experience has proved that if a clip of wool is honestly got up it only requires a few years of consistency to give the buyers confidence in the "brand," and that goes a long way towards selling the wool, and is of special value to the seller in time of depressed prices.

That careful classing is advantageous cannot be seriously doubted. In no line of business does slovenly work pay, and the wool trade is no exception to this rule.

If the fleeces are skirted and rolled up uniformly and neatly ; the bellies, pieces, and locks carefully picked over ; the stains dried before packing, and all the different "lots" baled up separately in bales of even weight, neatly sewn and branded, the general "get up" will be attractive, and will more favorably impress buyers at first sight than if the wool had been baled up in an unmethodical and careless manner.

The value of creating a good impression is very considerable, for in wool, as with any other produce, first impression goes a long way to help the sale and the Grower will get the benefit.

In view of what has been written, it is hard to understand why any wool is sent to market in such a deplorable state, and without any attempt to meet the buyers' requirements.

The argument "that it does not *pay* to class wool on the station or farm;" in fact, that it is "simply a waste of time and money to do so, because the wool has to be sorted over again when it gets to the factory," may seem very plausible, but it is only necessary to state that woolclassing on the station and wool-sorting in the factory are distinct branches of the wool trade, and the argument loses its effect.

Again, it is said " that the cost of classing is so great that any consequent increase in price will not sufficiently compensate the grower for money spent on the work."   Considering, however, that the extra cost of expert classing and general get-up rarely exceeds two shillings per bale, and experience proves that very often from one farthing to one halfpenny per pound, or seven shillings to fourteen shillings per bale more is paid for wool properly got up, it will be seen that there is no foundation for such a statement.

Wool Classing on Station,

Classing Small Graziers' Clips.

It is not the object of this book to deal at any great length with the larger grazier, but I think enough has been said in the last chapter to prove the necessity of woolclassing, and, seeing that the squatter will require to employ someone to do the work, I will not refer further to him under this heading than to state that there are now a number of young men in the country who have been trained at the School of Mines and Industries, and there are others being trained in that institution, so there should be no difficulty in obtaining the services of a competent wool classer. But, as it would not pay a small farmer to employ such a man, I will try and show as clearly as possible how he can class his own wool to advantage.

In the previous chapter it was pointed out that with a small farmer the same number of sorts is not necessary as in a larger clip, and I would further state that for a small farmer to attempt to make as many sorts as are made by a larger grower, would undoubtedly result in a loss. This statement is strengthened by the fact that small lots of three bales and under are sold separately from the larger ones, but as this matter has been clearly explained in the chapter on wool selling, I will not deal with it further here. Of course, it is quite impossible to lay down hard and fast rules as to how even small clips should be skirted or classed, as the conditions vary so much in different places, and thus it may be advisable to make one or two sorts less or more on some places than on others, but if the following hints as to how the smaller clips should be classed are acted upon, the result should be satisfactory.

MERINO.

Take the case of a small grower who does not run his sheep on the fallow, in good light country such as the hills district in the South, it will in most cases pay him to skirt his fleeces deeply when there is burr in the wool. The reason why it will pay to do so is on account of the frequency with which American buyers are now purchasing Adelaide wools which suit them. In order that this statement may be better understood, I would draw attention to what has been maintained in the previous chapter about the duty on wools imported into the United States of America, so that it must be clear that it would not pay the Americans to buy wool which would lose a large percentage in the process of manufacture.

There are also like conditioned clips among the hills in the lower North which may be treated in the same manner as those already referred to, but these are exceptions.

With the ordinary wool which is suitable for the English and Continental trade, it may not be advisable to skirt so deeply, because the skirtings or pieces in a small clip do not bring so much money as they do in a larger clip where there are a number of bales of this class of wool, whereas with a small clip there will be only a few bales.

Although I do not advocate deep skirting in these cases, I would strongly advise taking off all the very burry or dirty wool from the fleece, because if this kind of wool is left on the competition would be so restricted that the best price

would not be obtained.   Whatever skirting is done, care
and judgment must be used, so that wool will not be taken
off the fleece which should be left on, and at the same time
nothing should be left on which should be taken off.   This
evenness of skirting is a very important matter.

Referring to the subdivision of the fleeces, the greatest
care should be exercised in keeping out all the very fatty,
yolky ones, as, whether Americans are competing or not,
the presence of very fatty fleeces invariably interferes with
the sale of the clip.   In the case of the American buyers,
they don't want heavy, yolky fleeces at any price, and with
the other buyers the fact of the fatty wool, which is of less
value than the other, being mixed with the rest, makes it
impossible for them to arrive at the true value of the lot, and
it stands to reason that any doubt they have will be at the
expense of the grower.

Then again, if there are any fleeces which are much coarser
than the bulk, they too should be kept out, because many
buyers who are quite prepared to buy the ordinary wool, may
not be able to use the coarse.   Any very short or tender
fleeces should also be kept out.

As it is a general practice to shear the ewes, hoggets,
and wethers separately, it will be no trouble to bale these
different sorts by themselves.   Each of the sorts should
be baled separately and branded with some distinctive
mark, such as A for the best, B for the second, and C for
the coarse.   In some cases it might be advisable to put
the short and tender fleeces with the fatty ones, or it may
be best to put them in a bale or bag by themselves.

This baling of the hoggets, ewes, and wethers separately
does not necessarily mean an increased number of lots, be-

cause the grower when consigning the wool to his selling broker, leaves it to him to decide, after opening the wool, whether there is sufficient difference in type and value to justify him in keeping them separate when cataloguing for sale.   On the other hand, if the difference justifies it, he can catalogue them separately.

There is an advantage in baling the different lots referred to separately, even when they are all sold in one line, because, say there were four bales of A hoggets, five bales of A ewes, and six bales of A wethers, they would be catalogued in this manner :—

$$15 \text{ bales} \left\{ \begin{array}{l} 4 \text{ bales A hoggets} \\ 5 \text{ bales A ewes} \\ 6 \text{ bales A wethers,} \end{array} \right.$$

so that the buyer, having the exact number of each sort before him, would be better able to arrive at the true value.   All the B's could be sold together, as also could the C's.

Dealing with the skirtings or pieces, it is generally advisable to divide these into two sorts, which can be called " First Pieces " and " Pieces."   The first pieces to consist of all the bigger and cleaner wool, and the second to be made up of the trimmings or dirty edges taken from the bigger pieces.

Stained wool should be kept out and packed with the pizzle pieces, which should always be taken out of the wether bellies.

The belly wool and the pieces should be packed separately, and the bales should be branded what they really contain.

The floor and table locks can be packed together.

If the grower will supply the selling broker with full particulars as to the way in which his wool has been classed, it will materially assist him in his work of lotting the clip for sale.

The foregoing remarks, with slight modifications, will apply to the small grazier with up to several thousand sheep.

I might add that if a grower is in doubt as to whether his clip is one that will suit American requirements or not, his selling broker will be able to advise him.

SMALL CLIPS IN NORTHERN AREAS AND RIVER DISTRICTS.

Referring to the small clips in what is known as the outside country, the main conditions referred to in the previous chapter apply, but there is often one great distinction between the clips in question, such as the amount of sand found in the back of the fleece, and as a rule in these dry districts it will not be necessary to take off such a deep skirt. Where the clip is of a fair size and the backs very sandy, it is advisable to take the worst of the back out and bale that class of wool by itself. The reason for this will be clear when it is remembered that the sandy back wool is much heavier in condition, and thus of considerably less value than other parts of the fleece, and, besides, it is frequently excessively tender.

Country Class, School of Mines.

Classing Crossbreds.

Where Lincolns, Leicesters, or Romneys are used in crossing with the Merino, the wool produced will be more or less uneven in quality and length. This is only natural when one considers the great difference between the type of wool grown on the Merino and that on any one of the British breeds referred to.

In order to get anything like the best results from a clip of this kind, classing is absolutely necessary; in fact, even more so than in Merino. It is somewhat difficult to explain the different sorts without going into technicalities, but the following remarks should prove useful :—

In the first place, just how much skirt should be taken off must be left to the judgment of the grower, but I would point out that in the case of a very coarse breech being found on a comparatively fine fleece, it should certainly be taken off.

Referring to the fleeces, the untrained man should not attempt to make too many sorts. A fine, medium, and a coarse sort will be found in a general way to answer the purpose. Where there is a doubt in the minds of the grower as regards the quality of any particular fleeces, the length of the staple should be a guide. Thus, when making subdivisions on the lines referred to, a fleece may be thought just a shade too coarse for one of the finer sorts, but if it is short in staple, it may be safe to pack it with the higher grade. On the other hand, if it is long in staple, it will be better to put it down a sort. So important is the length of staple in the classing of crossbreds, that even if a fleece is on the fine side and of extra length, it will be wiser to class it with the coarser sorts. Any fleeces exceptionally fatty, matted, or discolored, should be kept apart from the other sorts.

Classing Lambs.

MERINO.—In classing lambs' wool, which does not hang together in the same way as the fleece of a grown sheep, it is necessary to pick the fleece up between two boards, and, as it is generally so short that it would fall between the spokes of the rolling table, it is also necessary to cover the table with some hessian or empty wool bales, the former for preference. The picker up should not put one fleece on top of another on the table, as this causes a lot of extra trouble in classing the wool.

If the lambs' wool is comparatively free from burrs or grass seeds, the bulk of the long light wool from the shoulder, back, &c., can be picked out and put in the best lot, whilst the shorter and heavier wool from which the dags and stains have been taken will make a sort of itself.

If some of the lambs' fleeces are burry, it is necessary to keep them apart from the free, because free lambs' wool generally sells remarkably well, whilst the burry is not so much sought after. Any fleece which is long and hairy might well be packed with the lower sort.

CROSSBRED.—As a rule the small owner has so very few crossbred lambs that it would not pay him to divide them up into quality sorts, and perhaps the best thing he can do is to take out all the lightest and best, the remainder, minus the dags, stains, &c., to be packed together. Any extra coarse to go with the shorter and heavier sort.

Classing Farmers' Clips.

Dealing with the wheat farmers clip, which is generally shorn from sheep running on fallow land, there is not a great deal of classing required. Where the ewes are all Merino, all that is needed is to trim or skirt off all the sweaty, dirty edges. The fleeces should then be rolled up from breech to shoulder, as is done by the graziers. It is only necessary to keep apart from the bulk of the fleeces any which are extra greasy and heavy, as well as the yellow discolored fleeces, and any which may be extra coarse. In a small clip the number of fleeces taken out might only be sufficient to fill a bag, but with a larger clip there might be enough to make a bale or even more. The trimmings or skirtings as a rule will be so short that they may be put with the locks, i.e., the trimmings from the legs, head, and face, as well as the small pieces which are found underneath the table. If, however, it has been found necessary to take off a little deeper skirt, naturally the trimmings will be better, and should be packed by themselves. In every case, belly wool should be kept and packed separate from the other portions of the fleece. Where ewes and wethers are kept, if possible, it is better to pack these types separately, although in all probability they will be sold in one line by the broker. Where Merinos and crossbreds are kept, the Merino wool must be packed entirely separate from the other. In fact, it is much better to shear the two lots separately so as to prevent any chance of mixing. In every case rams' wool should be kept by itself.

Before giving the reasons for the above, I might state that the table required is just the ordinary rolling table found in a wool shed, the top of which is made of spokes or broom handles about 1½ inches apart. The spaces allow any second cuts that are made to fall underneath.

The reason for keeping the heavy fleeces out is simply that they are as a rule of distinctly less value than the other wool, and, if left in, would materially affect the sale of the clip. The yellow and discolored fleeces are so unsightly that unless kept out, they too, would detract from the value. Then again the dirty, sweaty edges, being of much less value than the other portions of the fleece, are taken off so that the buyer may get as near as possible to the true value. The belly wool is kept separate, being of such different type that some buyers who want the fleece do not want the belly wool at all. Keeping ewes' and wethers' bellies wool apart does not amount to much, but where convenient it is as well to do so. It must be apparent that crossbred wool, being of different type and value to Merino, should be kept separate, so that buyers can get the wool they want without being compelled to buy wool which is of no use to them.

The reason for not advocating any further subdivision is that it would mean the making of many small lots, the difference in value of which would not justify doing it, and when it is remembered that small lots do not sell as well as big lots, the reason of this will be plain.

LAMBS.—These lambs, being generally run on fallow, need very little classing further than to keep the crossbred apart from the Merino, and all the daggy pieces out of each sort. However, in the event of there being a large lot of either Merino or crossbred, it might be well to further subdivide.

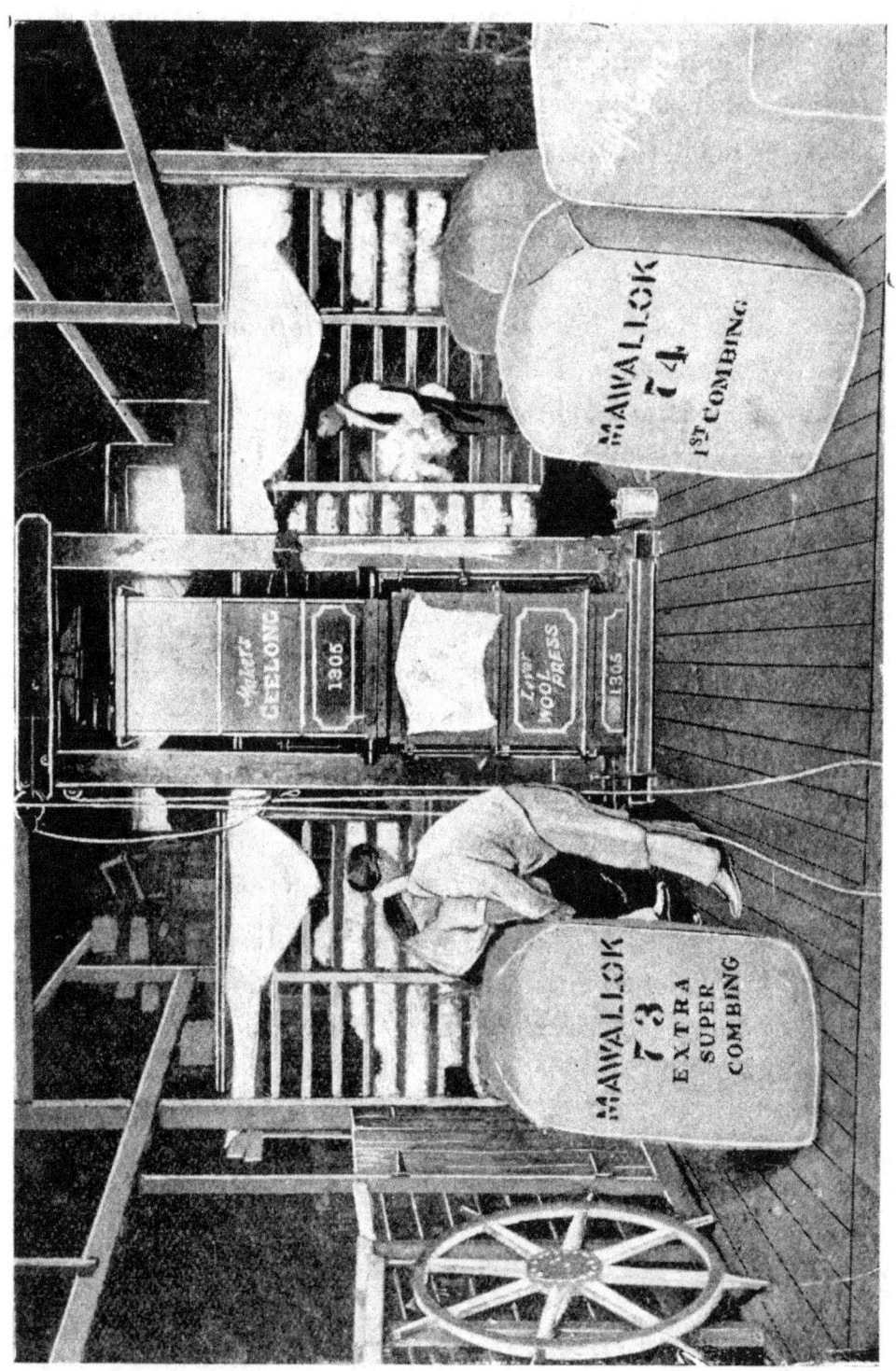

Wool Pressing.

Pressing and Branding.

———

Although it may be an advantage, it is not at all a necessity for the farmer to have a box press, more especially as the South Australian Government does not now allow a rebate of 10 per cent. for box-pressed bales carried on the railways, but care must be taken to pack the fleeces carefully and tread them down in such a way as to fill up the bale and make it look shapely. The bales should not be too heavy; 300 to 360 lb. is a fair weight.

The bales should be branded legibly on the top or flap end and on one side, say, with the name of the farm and initials of the owner, or at any rate some distinguishing mark, and they should be numbered consecutively from the first to the last bale of the clip. A stencil plate, which costs very little, makes a neater brand than when done with a stick or brush. Proper branding is of more importance than is generally considered, because it goes a long way towards preventing mistakes during transit to market, and when the wool is being taken into the store, also when it is being delivered after it is sold.

The following is an example which can be altered to suit owners' ideas :—

GJ

Pine Farm

7

A Ewes

DESCRIPTION OF WOOL.—That the wool should be properly described on the bale (the description is also printed in the sale catalogue) is a matter of some importance, as an injudicious description may lead to lessening the competition. As an example of how this may come about, take the terms "Firsts" and "Seconds," which are often applied to both fleece wool and pieces. These terms do not convey any meaning to the buyer as to the type of wool contained in either lot, because the firsts of one lot may not be as good as the seconds of another, owing to the fact that the wools have been grown in entirely different districts.

It frequently happens that a buyer with an order to purchase a good line of pieces, or a given quality of fleece, may find just exactly the wool he wants branded "Seconds." This description may lead him to pass it by without valuing it, because, to many principals the term "Seconds" suggests that something is wrong with the wool, or, at any rate, that it is not as good as what they require; they, therefore, make an extra careful examination of it in order, if possible, to discover something wrong, and everyone knows that whenever a fault is looked for it can generally be found, and, rather than court such special enquiry, lots branded "Seconds" are neglected by some buyers.

This is only one instance in which ill-advised branding of wool may tell against the grower, but it will be sufficient to show the importance of adopting some better system. The letters of the alphabet answer the purpose better than numbers, thus—Firsts can be branded A, Seconds B, and another sort C, or, if there are a greater number of lots, the best could be branded AA, the next A, other qualities BB and B, or as the owner wishes.

Wool Scouring.

The question is often asked, " Does it pay to scour such lots as short pieces, locks, stains, &c., before sending to market ?"   This is a very debatable question, and one on which much can be said on either side.   There is no doubt that where wool is grown a long distance away from market, and the cost of transit is great, that it does pay to scour the heavier lots, especially if there is a plentiful supply of good water available, but with the small farmer within easy access of the market, it is very questionable if the extra price which would be obtained for the scoured wool would make up for the amount of weight lost in scouring, together with the expense and trouble involved.

Of late years it has been noticed that foreign buyers have been purchasing locks and stained pieces in the grease, which proves that the competition has been general, but it should be pointed out that the lots purchased by the section of buyers referred to are generally station lots, and as a matter of fact the small farmers' lots are generally bought by the Australian wool scourers, who, of course, are speculators, and who scour the wool and offer it again in the scoured state.

This naturally opens up the question that if it pays these gentlemen to buy and scour, might it not pay the farmer to scour his own small lots ; but it has to be remembered that the buyers referred to purchase a great number of different lots, and, as they have every facility for scouring the wool as quickly and cheaply as possible, they have a distinct advantage over the small farmer scouring his one or two or more bales as the case may be.   Thus, generally speaking, I do not think the small farmer within reasonable distance of the market will gain anything by scouring his oddments, but where the cost of sending to market is anything over a penny a lb., and the conditions are favorable, it will more than likely pay to scour the heavier and earthier sorts.

In concluding this chapter I might just say that anyone scouring his own wool should be careful to use a good reliable soap, and one with as little free alkali as possible.

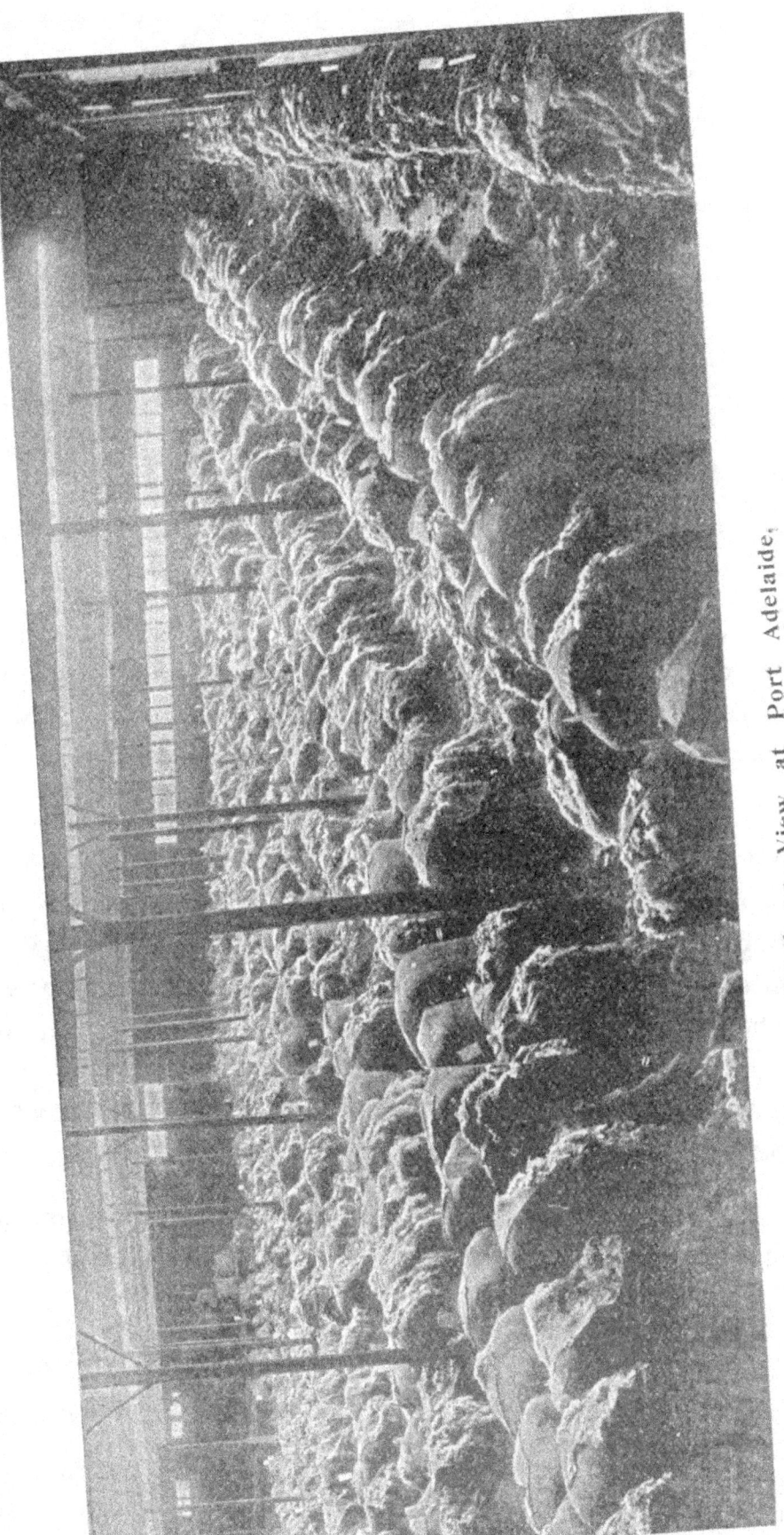

Wool on View at Port Adelaide.

Wool Selling.

In order to have a clear idea how the wool is dealt with by the selling broker after it has been received and weighed, it would be interesting for the farmer to visit the stores at Port Adelaide when his wool is ready for inspection prior to the sale, but, as this is not always convenient, the following will give a very good idea of how the wool is handled.

On account of the large quantity of wool offered on the one day, it is impossible to show every bale, and buyers are quite content with the following proportions being shown, viz. :—

In lots of 10 to 100 bales and over, 20 to 25 per cent. ; in lots of 5 to 10 bales, 25 to 30 per cent., whilst in lots of four bales and under, every bale must be shown. These proportions apply to Merino wools, but with crossbred wools it is often necessary to show as many as 50 per cent. of the bales. The sample bales are then laid down in rows and the tops cut open, and where any suspicion of uneven packing is apparent, the bales can be cut about in any way, so as to show the wool as it really is.

The selling brokers supply buyers with a catalogue (a sample page of which is shown at end of this chapter) of all the wools offered for a given sale.

Each lot has a number assigned to it, and opposite the number is given the brand and description of the wool, with the number of bales in the lot.

Some time before the sale, buyers who have been provided with the catalogues inspect each lot so as to find those suitable to their requirements, and it is interesting to see the way in which they handle the wool. No amount of trouble is too great in order to ascertain its true value, and the wool store presents a sorry picture when the buyers have finished valuing. Wool which has been pulled out of the bales is sometimes heaped up two or three feet high. This wool is repacked into the bales after it has been sold.

Briefly, the principle on which buyers work is as follows : —First, they find the kind of wool they require, and proceed to estimate what percentage it will yield when scoured, *i.e.*, how many pounds of clean wool each hundred pounds of greasy wool will return after it has been scoured. Of course, they know the price they can pay for the wool when it is clean scoured, as this is given them by their respective firms, and, if they can arrive at the true yield, it is only a simple matter of calculation to ascertain what they can pay for the wool in the grease.

For instance, a buyer has instructions to buy wool at, say, 22 pence per pound clean scoured. If the wool will yield, when free from all impurities, 50 per cent. or half weight, he can pay 11d. per lb. in the grease, so that, having found the quality of the wool he requires, which is a comparatively simple matter to the trained expert, the main point is to estimate the quantity of foreign matter in it, in the shape of yolk, earth, burrs, &c.

To show the importance of this, I would point out that when wool is sold at over 9d. per lb. in the grease, it will be found that one per cent. more or less foreign matter makes a difference of a farthing a lb. in the price of the wool in the grease. Thus, take a clip which, when scoured will yield, say, 40 per cent. and is worth 9d., if the same kind of wool will yield 41 per cent. it is worth 9¼d., or if it would yield 44 per cent. it is worth 10d. This is somewhat difficult for the layman to grasp, as 2 per cent. or 3 per cent. difference in yield is not always apparent to the untrained eye, and this fact often accounts for the dissatisfaction of some growers with the price obtained by their selling brokers, and they cannot understand why their neighbor who has exactly similar sheep as they have, should get from ¼d. upwards more than they got. The real reason in most cases is that their wool, being perhaps better nourished, contains more yolk, and, although they do not get so much per pound as their neighbor, they may in reality be getting a considerably higher return per sheep.

When the buyer has arrived at the price he is prepared to pay, he writes that down in his catalogue alongside the lot he has been valuing, and when the wool is offered at auction in Adelaide, he is guided entirely by his catalogue.

While the buyer is busy the selling broker is not idle, for he, too, values every lot, so as to protect the grower, and this valuation guides him when selling the wool at auction, and, unless growers' reserves are higher than his valuation, he will sell the wool catalogued according to the valuation he has placed upon it.

The wool sales in Adelaide (or anywhere else for that matter) are always an interesting sight, but it requires a more fluent pen than mine to aptly describe them. Suffice it to say that " Bedlam," " Pandemonium," and other words of that nature are terms which are very frequently applied to a wool sale by those who see it for the first time.

During the progress of the sale it will be noticed that each broker sells all the lots printed in the larger figures first, and after all the brokers have finished selling these lots, the first seller starts again with the lots printed in smaller figures. These latter are called " star " lots, and are composed, as will be seen by the sample page of wool catalogue, of lots of three bales and under. The arrangement referred to has come into vogue in order to assist buyers in getting through their work in a reasonable time. Hence, when the larger lots have been sold, the principal buyers leave the room, and if they have valued the small lots at all, they leave their prices in the hands of a buying broker. In the light of this, it must be evident that the smaller lots do not receive the same competition as the larger ones, and they do not as a rule bring the same price, even when of the same quality and value. This, then, is the reason why I do not advocate classing the farmers' clip into too many lots, for, where it can reasonably be avoided, it is better not to have lots of less than four bales. It might be suggested that all the lots should be sold in rotation, but the reason for not adopting the practice will be clear when it is pointed out that the large and important buyers, who so materially assist the market, are as a rule too busy to devote themselves to the small lines, and if the larger lots were not sold separately from the others, we would probably have less competition all round. This system is in vogue in all the other States and in London, as well as in Adelaide.

There is often a misunderstanding amongst farmers as to the exact meaning of the terms " tare " and " draft " used in connection with the sale of wool.

Tare means the allowance made for the weight of the woolpack, and in this market is fixed at 11 lb. per bale.

Draft is an allowance to buyers of 1 lb. per cwt., and is a custom which dates back to the earliest history of the wool trade.

Sale No. 5—26th November.

Lot.	Mark.	Greasy.	Bales.
301	H M D GLENROY	4—Hgts. 3—Wrs. 4—Ewes	**11**
302		Lambs	3
303		Bellies	3
304		Pcs.	1
305		Locks	1
306	B S L HALLETT	2—H 2—W	**4**
307		Ewes	3
308		Lbs.	1
309		Bellies	2
310		Pcs.	1
311	G D FINKLEY	AA E H	**17**
312		AA E	**16**
313		A E H	**5**
314		A E	**35**
315		A W	**36**
316		2 - BB W H 3—BB E H	**5**
317		BB E	**18**
318		BB W	**4**
319		17—CC EH 18—CC E 1—CC W	**36**
320		AA Lambs	**18**
321		A Lambs	**10**

Wool Sale.

Wool for Show and Competition.

———

I have always been of opinion that it would be well to give greater encouragement to the exhibiting of wool at many of our country shows, and in order that farmers might have a true idea as to how the wool is judged, and thus be encouraged to compete with one another to their mutual advantage, the following has been written :—

Supposing the competition be in the Merino class, it frequently happens that one farmer selects a fleece of the " finest " quality, another the very " brightest " looking fleece, a third a fleece with the "waves" or "crimps" well and evenly defined, a fourth considers the "weight" of the fleece to be the main point.

In each case the exhibitor, in making his selection, may have been influenced by one special qualification, and perhaps has entirely overlooked excellent qualities which many other fleeces in his clip may possess.

When the prize is awarded it may so happen that the "finest" fleece has been placed first, or that the "brightest" has secured highest honors.

In either case it is quite possible that the owner of the prize fleece, as well as others interested, may be carried away with the idea that the winning fleece was selected by the judge solely on account of its "fineness" or "brightness," as the case may be.

As a matter of fact the "fineness" or "brightness" may have had only comparatively little to do in influencing the judge in his decision, because, generally speaking, the relative difference in the fineness or coarseness of Merino wool, grown in the same locality, is not sufficient to very materially affect the value of the fleece. Further, as the brightness of the fleece does not very materially add to its value, it will be seen that that qualification is not a very important factor, and when it is stated that well-defined crimps, or any other characteristics, do not so materially increase the value of the wool, the question may be asked, " What then is the main point to be kept in view when selecting Merino wool for show competition ?" and here the principle of wool judging may be explained.

It consists, practically, in placing everything else in subordination to the cash value of the exhibit, and, seeing that general characteristics play a rather unimportant part in this respect, it must be clear that the percentage of clean wool is the principal factor.

The cash value of the exhibit is also the determining factor with the judging of " other than Merino " wools, but, seeing that the relative difference in fineness or coarseness in this class may be so great as to affect the price per pound very considerably, this qualification is taken much more into consideration, along with the percentage of clean wool.

Take, for example, a competition in the "other than Merino" class in which there are four competitors for a prize offered for the best fleece.

A sends in a typical Dorset-Merino cross fleece.

B exhibits an ordinary Leicester-Merino fleece.

C pins his faith to a Shropshire cross fleece.

D, who is famed for his good Lincolns, enters the lists with his most lustrous fleece.

The fleeces are laid out ready for the judge, who, without the least sentiment concerning the particular breed of the sheep from which the fleece is taken, proceeds with his work.

If the fleeces have already been weighed and ticketed, so much the better, if not this is the first thing to be done. Having the weight of each fleece before him, the judge begins with No. 1 to ascertain the value per pound, and places that against the weight, thus :—

Lot 1.   1 Fleece, weight 12 pounds, value 9d. per lb., equals 108 pence.

Lot 2.   1 Fleece, weight 11 pounds, value 11d. per lb., equals 121 pence.

Lot 3.   1 Fleece, weight 12 pounds, value 9d. per lb., equals 108 pence.

Lot 4.   1 Fleece, weight 16 pounds, value 7½d. per lb., equals 120 pence.

From this it will be seen that number 2 will be awarded the prize, not because of its particular type, but simply on account of the monetary value of the fleece.

This example is, of course, exactly typical of the method in which *Merino* wool also is judged, hence there is no necessity to give a detailed example of how the judging is carried on in this class.

As the object of Agricultural Societies in giving prizes for wool competition, is to encourage woolgrowers to keep sheep which will be most profitable to them, and not to foster the growing of any fancy wool, it will be clear that the judging could not satisfactorily be done in any other way than that indicated.

Wool Judging.

Sheepskins.

There is not the slightest doubt that even at the present time a large amount of money is annually lost to sheep farmers (which might easily be saved) through the lack of knowledge or negligence in connection with the getting up of skins for market.

Certainly it must be admitted that of late years much more care has been taken than formerly in this respect, but there still remain many points which, if attended to, will further considerably lessen the loss, and the following suggestions, if carefully acted upon, will prove of great advantage in this respect.

It should be understood that, although skins are sold at so much per skin, the buyer really values them at so much per lb. To make this absolutely clear, I would state that skins are put up for sale in heaps, according to their quality and weight, and the selling broker carefully weighs each skin, so that the buyer in his catalogue gets to a fraction the average weight of the skins in the heap. After valuing the heap at so much per lb., the buyer simply multiplies the weight by the price per lb. he is prepared to pay, and thus he arrives at the price he can give for each skin.

For example, if a heap of skins average 10½ lb. and the value is 8d. per lb., the buyer can pay 84d., or seven shillings each. This being the case, it is advisable, when the sheep is killed, to completely strip the carcase, including cheek, poll, legs, and tail, as every extra pound of weight increases the value of the skin.

The killing should be done, if possible, on a draining board, or, at any rate, some clean place. Care should be taken to prevent the blood getting on to the neck of the skin, otherwise the wool will look unsightly.

It is advisable to skin the sheep almost immediately after it is killed, because unless this is done, blood will collect in the veins, and in a measure stain the pelt.

Cutting the skin should be carefully avoided, because by doing so the sale value is considerably lessened. In order to avoid cutting, it is well in removing the pelt to use the hand, more than the knife, as by this method there will also be less chance of leaving lumps of fat on the pelt.

Painting.

Very soon after the skin has been removed it should be painted with an " anti-weevil " wash in order to prevent the weevils from getting a hold on the skins. Weevil-eaten skins are of even less value than those which have been cut. In fact, when very " weevily," not only is the pelt actually destroyed, but the wool is also damaged, for, in order to free it from the damaged pelt, it must be (to use a fellmongers' term) " pied "—that is, thrown on a heap to be sweated, with the result that the wool generally becomes more or less discolored.

A good anti-weevil wash, and one that can be thoroughly recommended, is the following :—" Put eight gallons of water into a copper with 10 lb. of soda ash, ½ lb. of Barbadoes aloes ;

stir well until boiling, then add 15 lb. of arsenic." The mixture will rise like boiling milk. To prevent boiling over it is well to have a bucket of cold water handy to pour into the copper.

One part of the mixture to five of water is about the right proportion for painting. As this paint is poisonous, great care must be taken. It cannot be too strongly emphasised that, in painting, every part of the pelt must be touched, and the wash should not be applied until the pelt is somewhat firm.

A prepared "weevil-wash" can be procured at a moderate cost, and where small quantities are required, perhaps the better way is to purchase some reliable mixture rather than make it up.

DRYING.

The work of drying is of more importance than is generally imagined, because, should the skins be improperly dried, they are apt to become unshapely and wrinkled, with the result that they may not sell to the best advantage.

In order to obtain the best results, the skins must be dried in the shade, and not exposed to the sun and weather ; hence. where it is at all possible. they should be dried under cover.

Some time ago drying on a frame was in common practice, but the greatest advantage derived from this system was that it led people who had been hitherto careless to be more methodical. For practical purposes all that is necessary is to fold the skins lengthways, from head to tail, over a beam or rail, taking care that the edges of the skin are kept straight.

It is also very important not to have one skin touching another until they are dry. When dry, the skins should be taken from the beam and stacked one on top of the other, pelt to pelt and wool to wool, as this will prevent the pelt from getting too dry, when it would be liable to crack.

In packing the skins for market it is most important to see that they are thoroughly dry, as otherwise they may get mildewed, and, if left long enough in the bundle, may rot, so that not only will the damp skins be valueless, but they will affect the others in the bundle which may have been perfectly dry.

In bundling the skins to send away, the outer or woolly side of the skins should be exposed, as there is much less chance of the skin being injured on the woolly side than on the pelt side. A piece of bagging at each end before the wire is put round, will also further protect the skins. It is not wise to make a bundle too heavy, as it causes inconvenience in handling. About two to two and a half hundredweight is a fair weight for a bundle of skins.

It is needless to add that where convenient it is well to send the skins to market as frequently as possible, as, besides having them out of the way and getting the cash more quickly, it stands to reason that they do not improve by being left about.

Explanation of Terms Used or Known in South Australia.

———

BURRY—Wool containing burrs.

BULKY—Solid, a wool which fills the hand.

BROAD STAPLE— A large number of fibres grouped together in clusters or locks, sometimes called a " Big Lock."

BRIGHT—Wool of good color, attractive, generally applied to Merino.

BROKEN—Specially good pieces made where deep skirting has been carried out.

CHARACTER—The crimps or undulations as seen in the wool by the naked eye.

CLAMMY—Wool which has a sticky feeling through not being properly scoured.

COMBING—Applied to wool of a fair length of staple as against the very short wool.

CLOTHING—The very short wool in the clip.

CONDITION—The amount of yolk or foreign matter in the wool.

COTTED OR MATTED—Wool that has felted while on the sheep's back.

CROSSBREDS—Wool from other than pure bred sheep.

CRUTCHINGS—Wool trimmed from the hind quarters before shearing.

DENSITY—The closeness with which the fibres are grown together.

DEAD WOOL—Wool taken from a sheep which has died.

EXPLANATION OF TERMS.

DRY WOOL—Term applied to scoured wool which is harsh, generally a sign of being over-dried or over scoured.

ELASTICITY—The power that the fibre possesses of returning to its original length after being stretched.

EWES—Wool from the ewes.

EARTHY—Wool containing a large quantity of earth.

FINE—Thin fibre, or wool, the fibre of which has a small diameter.

FIBRE—The individual strand of wool.

FLUFFY—Light, open, downy appearance.

FRIBBY—Fleece with the second cuts adhering to it.

FATTY—Wool in which the grease is clogged and sticky.

HOGGETS—Wool from one year old or two-tooth sheep.

HARSH—Wool with a lack of yolk and which handles unkindly.

KEMP—Small straight hairs, white or black, hard or brittle, generally found on the top of the head and legs.

LAMBS—Wool from the lambs.

LOCKS—Trimmings from the legs, face, &c., as well as the second cuts and fribby pieces which fall underneath the rolling table.

LENGTH—Referring to the length of the staple.

LEAN STAPLE—Small number of fibres grouped together in clusters or locks, sometimes called a " Thin lock."

LUSTROUS—Possessing a good sheen or gloss, generally applied to the long wool of such breeds as the Lincoln, Leicester, or Romney.

MOITY—Wool containing all kinds of vegetable matter in the shape of hay, straw, twigs, leaves, seeds, &c.

MUSHY—Perished, wasty.

NOILY—Wool with a fuzzy or mushy tip.

PLAIN—Wool lacking character.

QUALITY—Generally refers to the fineness or coarseness—that is, the thickness or thinness of the fibre, although it sometimes refers to the general characteristics of the wool as a whole.

ROBUST—Wool strong, or thick of fibre, applied to Merino.

RAMS—Wool from the rams.

ROPEY—Wool which has been twisted or partly felted in scouring.

SERRATIONS—Serrated or saw-like edges of the wool fibre as seen under a microscope.

STAPLE—Group of fibres clustered together, commonly called a lock.

SKIRTY—Fleece which has not been properly skirted.

SEEDY—Wool with grass or other seeds in it.

SHANKINGS—Trimmings from the legs which generally show kempy hairs.

SHAFTY—A big wool of good length of staple.

SLIPE—Wool which has been pulled from the skin after the use of chemicals, lime, or by sweating.

STUMPY—Wool with a short staple and broad tip.

TIPPY—Wool with an extra amount of grease at the tip.

TENDER—Weak, not necessarily in one part of the staple only.

UNSOUND—Wool with a break generally caused by a lack of feed during some period of its growth.

WASHED FLEECE—Wool which has been washed on the sheep's back before shearing. (This type of wool is now almost out of date, although 20 or 30 years ago many of the best flocks were washed before being shorn).

WETHERS — Wool from male sheep which have been castrated.

YOLK—Natural suint or grease exuding from the skin of the sheep, which acts as a lubricant to the fibre.

YIELD—Referring to the actual weight of clean wool after being scoured.

Terms Used by Manufacturers.

NOIL—The short fribby pieces which have been combed out of the tops.

TOPS—Wool which has undergone part of the process in the manufacture of worsted goods—that is, the wool has been scoured, and after undergoing minor details it is combed out into a rope-like strand. During the process the fibres have all been straightened out and laid paral'el and all short fribby pieces combed out. The process is something like a lady combing her hair.

60s.—(Technical term used in the worsted trade), meaning 60 counts. The term when applied to raw wool means its capacity to be spun into a weavable thread of a given length and weight. In yarn or thread the system refers to the number of hanks in a lb. weight, thus 60s. means 60 hanks of yarn, each measuring 560 yards in length, or 33,600 yards of spun yarn to the lb. weight.

56s.—Means 56 counts, or 56 hanks, each measuring 560 yards to the lb. weight.

Referring to coarser sorts such as Lincolns, which are generally known as 36s., which means 36 hanks each measuring 560 yards to the lb. weight.

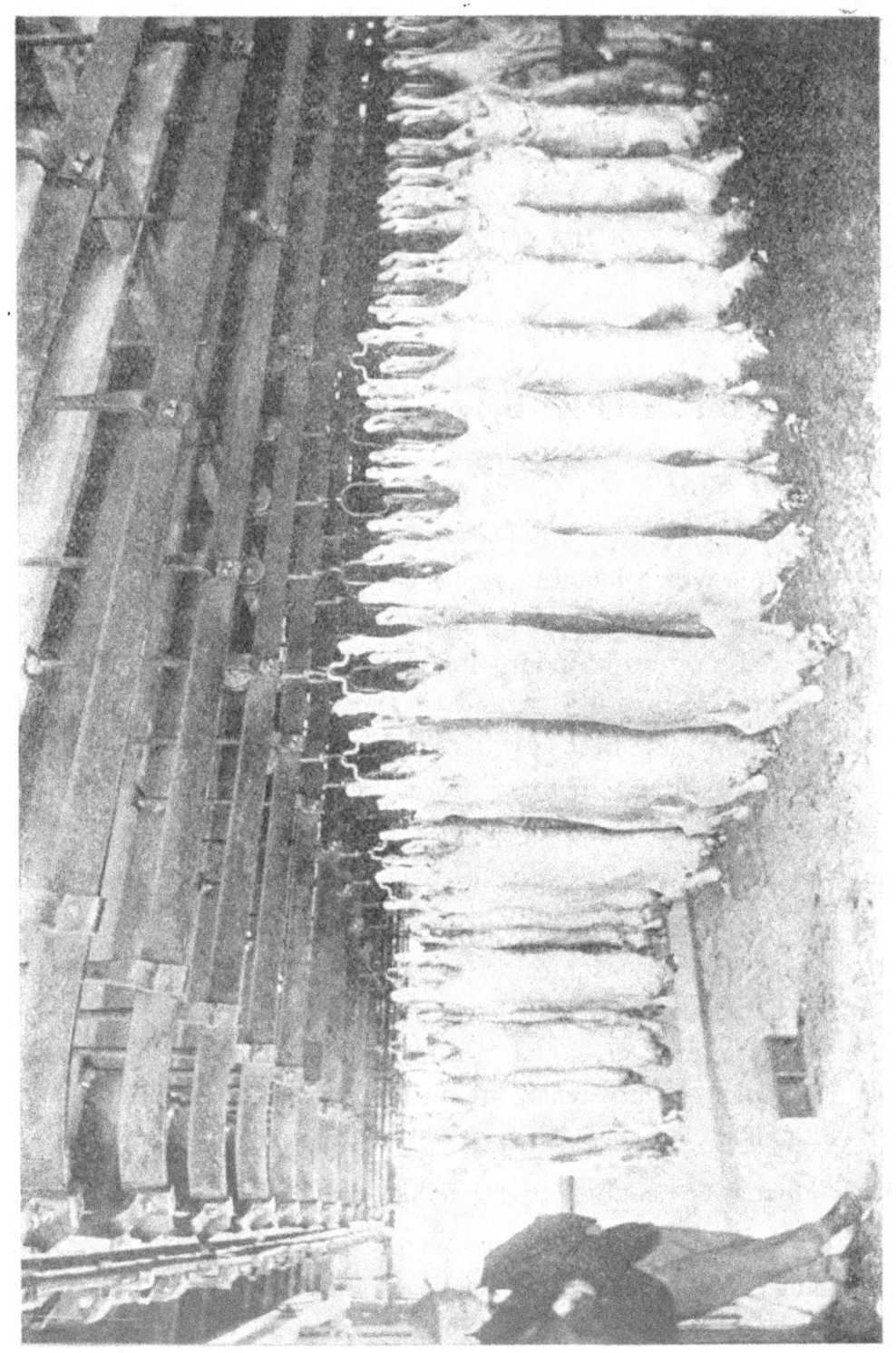

Freezing Chamber, Government Freezing Works, Port Adelaide,

Frozen Lamb Trade.

A RECORD SEASON.

The manager of the Government Produce Depot has forwarded to the Commissioner of Crown Lands the following report on the lambs and mutton exported during the 1906-7 season :—

QUANTITY.

So far as numbers go, the frozen lamb season in South Australia, which closed in February, was a record one, for more carcases of lamb were treated than in any previous season.   Taking both Government and private shipments, there was an increase in the exports of over 14,000 carcases, as compared with the previous year's figures.   In view of the adverse season for raising prime lambs, the inference is that farmers and pastoralists are awakening to the advantages of the frozen lamb industry, and in a more favorable year it may be expected that the facilities for slaughtering and freezing will be more severely taxed than they were in the season that has just closed.   The total of South Australia's exports of lambs and mutton is shown in the subjoined table, the number of carcases sent through the depot and those frozen privately being shown separately :—

Year.	LAMBS.			MUTTON.			Season's Total.
	Through Depot.	Privately.	Total.	Through Depot.	Privately.	Total.	
1895-6	1,751	—	1,751	1,097	—	1,097	2,848
1896-7	10,606	—	10,606	675	—	675	11,281
1897-8	3,534	—	3,534	463	—	463	3,997
1898-9	38,620	—	38,620	2,052	—	2,052	40,672
1899-1900	89,980	—	89,980	1,334	—	1,334	91,314
1900-1901	94,597	—	94,597	7,122	—	7,122	101,719
1901-2	45,440	47,134	92,574	—	—	—	92,574
1902-3	63,798	53,045	116,843	19,464	18,656	38,120	154,963
1903-4	64,939	91,436	156,366	10,521	10,390	20,911	177,277
1904-5	155,858	37,882	193,740	2,254	311	2,565	196,305
1905-6	161,066	66,317	227,383	—	—	—	227,383
1906-7	163,810	87,750	251,569	2,613	434	3,047	254,616
	898,999	383,564	1,277,563	47,595	29,791	77,386	1,354,949

QUALITY.

Notwithstanding the satisfactory growth in the industry from the point of view of numbers, the season has been a most disappointing one, and the quality of the lambs offering was far below the average of previous seasons. The figures given below show that more lambs than usual were rejected and a larger percentage of second-class lambs was sent home. In consequence there has been some complaint from London. The lambing for the season was particularly good, being at least 10 per cent. better than for the previous season ; but no other advantage can be claimed. The weather conditions were very unfavorable. The farmers who went in for early lambs found that when the lambs were dropped the feed was still dry. As the year progressed a very wet winter set in, and the reports that came to hand showed that the continuous wet weather, without a fair proportion of sunshine, was materially affecting the maturing of the lambs. The result was a very late season, and lambs that should have been exported in October were not advanced enough until November. The extra month in the paddocks developed a larger frame, but there was almost an entire absence of " sap " and that shapeliness of carcase which is so essential in a first-class export lamb. The end of the season, so far as the meteorological conditions were concerned, was as unsatisfactory as the beginning. After the late and continuous rains there came a very sudden spell of hot weather in October. One hot Sunday, particularly, did incalculable mischief. From all parts of the country reports came to hand that the grass had been almost dried up, and the farmers found themselves in an unenviable position. They had thousands of lambs wanting three or four pounds more to top them off, and while they now had suitable weather their grass was gone or drying fast, and the seed difficulty was presenting itself. It was at this period that the depot began to feel the peculiar conditions of the season. Arrangements had been made, and the necessary labor obtained to start in earnest from the beginning of October, but the continuous wet weather kept the lambs back. Then the hot spell came, and matters developed with a rush, and the depot was taxed as it never had been before. From October 18 to November 24 the depot was over-supplied, and during the five and a-half weeks 113,030 lambs, or two-third's of the season's supply, were passed through. Even this did not wholly relieve the position, and the department was almost bombarded with requests to try to squeeze in more, the chief reason given being that the lambs were going back on account of the grass seeds. It will readily be seen from what has been said that the depot was flooded with immature lambs, and that this was caused by the late season, the rapid drying up of the feed, and the abundance of grass seeds.   This is shown by the following comparison of average weights :

Year.			Depot.	Private Works.
			lb.	lb.
1901-2	..	..	33.13	—
1902-3	..	..	32.03	—
1903-4	..	..	34.97	—
1904-5	..	..	35.22	—
1905-6	..	..	37.07	35.57
1906-7	..	..	33.70	33.40

The improvement in the averages between 1901-2 and 1905-6 is a fair indication of the strides being made in the breeding of suitable lambs for export.   The number of lambs averaging 40 lb. or over was 18,680 in 1904-5, 29,420 in 1905-6, and only 2,653 in 1906-7.

"Rejects" and Grading.

The season has brought to a head the difficulty experienced with those lambs which, after being slaughtered, have been found unfit for export, and which bear the trade name of " rejects."   This year there were 4,491 lambs rejected after slaughtering, as well as 1,657 on delivery alive, or a total of 6,148.   The Adelaide saleyards are, however, a much better index of the number of unsuitable lambs offering. Between October 3 and December 19 approximately 112,247 lambs were yarded at the saleyards, and only 52,772 or less than half, were purchased for export.   It is making a low estimate to say that 50,000 lambs, which should have been export lambs, were not frozen and sent away because, by reason of the unfavorable season, they were not fit.   Although there were 2,700 more lambs exported than in the previous season, the number of first-class quality was less by 20,171 carcases and the number of second quality increased by 21,753. In addition, 1,961 carcases were for the first time placed in a third class, as was done in the previous year by private exporters. The following figures for the last three years will show how the lambs exported through the depot and privately were dealt with, as well as the number of rejects :—

DEPOT.

	First.	Second.	Third.	Rejects.
1904-5	135,868	21,657	—	1,796
1905-6	114,426	45,850	—	2,359
1906-7	94,255	67,603	1,961	4,491*

PRIVATE.

	First.	Second.	Third.	Rejects.
1904-5	—	—	—	—
1905-6	51,287	12,108	2,229	609
1906-7	20,484	46,871	19,210	2,387

The percentage in each class and the percentage of rejects to the whole are as follows :—

DEPOT.

	First.	Second.	Third.	Rejects.
1904-5	85.28	13.59	—	1.12
1905-6	70.35	28.19	—	1.45
1906-7	56.01	40.17	1.16	2.66

* 1,657 rejected alive.

PRIVATE.

	First.	Second.	Third.	Rejects.
1904-5 ..	—	—	—	—
1905-6 ..	77.43	18.28	3.36	0.90
1906-7 ..	23.02	55.69	21.59	2.68

There was practically the same percentage of rejects at the depot as at private works. The above figures tell the tale of the season in unmistakable terms. Instead of 85.28 per cent. of first-class lambs, as in 1904-5, or 70 per cent. as in 1905-6, only 56 per cent. were able to pass muster for inclusion in the first-class this year. The figures from private works are probably more really descriptive of the ruinous state of the weather conditions, for the percentage of first-class lambs dropped from 77 per cent. in 1905-6 to 22 per cent. in 1906-7. During the lamb season the saleyards throughout the State were heavily stocked with store lambs, for which breeders had to accept prices, ranging from 6s. to, perhaps, 8s. per head, according to whether they were shorn or in the wool. Against this it will be seen from the following table that the average net sales of lambs shipped to London this season is approximately 13s., including the value of skins and fat. Making all allowances for the season, there is a lesson to be learned somewhere. New Zealand overcame a similar difficulty by artificially feeding lambs, which could not be fattened before weaning. While showing above that there is a clear loss of 5s. to 6s. per head on those lambs which export buyers would not look at, there is another and more serious loss. A perusal of the season's grading explains my meaning, as second and third grade lambs are worth, on an average, about one-half penny per lb. less than first-grades, and on this showing alone producers are losing thousands of pounds. Taking this season's average at a fair estimate of 3 lb. per carcase, less than that of last year, there is a loss of over £12,000 to the industry. The particular point to be emphasised, however, is the loss in reputation through shipping lambs of inferior quality ; and, taking New Zealand prices as a standard value, this State, by building up an equal reputation, would increase the normal value of the lambs by at least 2s. per head, as we are now accepting for our best quality nearly 1d. per lb. less than " prime Canterbury." Experienced men with a thorough knowledge of the meat trade of the world have pronounced our primest lambs to be equal to the very best placed on the English market, with this special advantage that they are milk-fattened. It is, therefore, worth while making a definite effort to bring our average up to somewhere near our present best. The question as to how this can be done is answered by the fact that it is not an extremely difficult matter to grow artificial feed that would relieve the necessity of depending altogether on natural grasses, which, through weather conditions, may be unobtainable at critical times. The main points are to first breed a suitable lamb, and then see that it gets no check from the time it is dropped until slaughtered for export. Producers who are endeavoring to improve the reputation of South Australian lambs must be protected by a rigid rejection of all inferior carcases offered for shipment. It is a difficult matter to decide on a "standard " for shipment, but against the strong objection some may make to fixing a "standard for export," it can be pointed out that Weddle & Co., who are one of the biggest distributing houses of Australian meat in England, strongly advocate a fixed and uniform standard throughout Australia.

DRESSING.

At the end of the season the depot was subjected to some

criticism on account of the dressing of the lambs ; but both in London and at this end, it was agreed that the criticism was not justified.   What was entirely due to the peculiar season was erroneously ascribed to faulty dressing.   It is a well-recognised axiom in the frozen meat trade that proper dressing will not make inferior lambs first-class, but inferior dressing will make good lambs second-class.   Exporters, generally, expect their lambs to be so dressed as to be equal to those of New Zealand, but under the special condition that prevailed last season this was impossible.   Even in a good season New Zealand and South Australian lambs are vastly different to dress.   The former are topped off on rape and turnips, and are trucked in almost every instance but a short distance.   Before being slaughtered they are carefully live graded, and those that fail to pass the grader can be turned out into convenient green paddocks until they are fat.   New Zealand is practically free from the hot spells that South Australia is subject to at the most critical time.

In a letter, dated February 21, 1907, the Commercial Agent wrote :—" While I do not wish to say that the dressing of our lambs this season is altogether faultless (neither is New Zealand), I do most emphatically say and I can produce overwhelming evidence to substantiate my statement, that it is not the dressing, but the plain quality of the lambs that is responsible for the low price, coupled with the extensive quantities that came on the market almost at one time.   I have not heard a single complaint about the dressing of any lambs that bore evidence that at the time they were killed they were in good condition ; and the highly satisfactory prices that have been obtained for these. to my mind, is convincing.   As previously reported, the majority of experienced buyers here could see for themselves that the lambs never properly matured. The fact that so many carcasses have a frame that should carry 10 lb. more meat than they do, and the absence of fat, are enough to satisfy any practical man that it would be impossible to dress them equally with the matured sappy lambs, such as most of the New Zealand lambs are. It is quite erroneous to imagine that no badly dressed lambs come on to this market from New Zealand.

The Agent-General, writing on February 22, gives an account of a visit paid by him to the Nelson wharf stores, where South Australian lambs, "some of good quality and in excellent condition, and others in poor condition," were being unshipped. In the course of his letter he says :—"It is useless to say that the fault with these lambs (those in poor condition) is either in dressing or shipping, because it was easily observable to anyone that their frames were sufficiently large for lambs of 40 lb. weight, while their weight would probably be about only from 34 lb. to 36 lb. The English purchasers recognise at once that there is just as much bone in a lamb of this kind as there would be provided the weight was 40 lb., and that all the loss in weight is a loss of meat, which makes material difference in the value of the article on the market." Mr. Jenkins goes on to point out the necessity of sending home only the best in order to establish a reputation similar to that which brings such good prices for New Zealand products.

THE LONDON MARKET.

It will be seen from the table of prices for each shipment made during the last season that the market in England was very high at first, and that it has been gradually falling. This was due to a scarcity of lambs between the end of the New

Zealand season and the beginning of the Australian supplies.
Shippers who were fortunate enough to have even a few lambs
in London at the critical time were able to obtain prices as
high as 6½d. per lb. gross; but very few arrived, and the
market price gradually dropped as the supplies were landed.
Growers who shipped early on their own account realised much
better prices than were offered locally, but the Adelaide
market gradually hardened in sympathy with London, and
those who shipped late, calculating on the London market
remaining high, will be disappointed with their final results."

SUMMARY OF LAMB ACCOUNT SALES TO DATE.

Boat.	No. of Lambs.	Gross returns, London.	All Charges.	Net* return, Adelaide.	Date of Shipment.	Date of Sale in London.	Date of Payment in Adelaide.	Gross Prices per lb. Average of each Shipment.		
								1 gr.	2 gr.	3 gr.
					1906.	1906.	1907.			
Ayrshire	100	14/1	3/6	10/7	Oct. 6	Nov. 28 to Dec. 1	Jan. 11	5.32d.		
Macedonia	91	14/8	3/8	11/0	Sept. 6	Nov. 15	Jan. 7	5.25d.		
China	206	13/7	3/6	10/1	Sept. 20	Nov. 5 to 15	Jan. 7	5.16d.		
						1907.				
Morayshire	2,917	13/1	3/7	9/6	Nov. 5	Jan. 7 to 23	Mar. 1	4.90d.	4.50d.	4.25d.
Somerset	1,160	13/3	3/9	9/6	Nov. 10	Jan. 4 to 26	Mar. 6	4.73d.	4.29d.	4.49d.
Essex	986	12/4	3/9	8/7	Nov. 19	Jan. 29 to Feb. 6	Mar. 23	4.31d.	4.24d.	4.07d.

*Add values of skins and fat, approximately 4/6 per carcase.

Cables from London give later shipments as follows:—

Sailing date.				Cable price.	
December 5, 1906	..	..	Indradevi	First grade, 4½d.	Second grade, 4d.
December 26, 1906	..	..	Fifeshire	First grade, 4¼d.	Second grade, 4d.
January 8, 1907	..	..	Dorset	Estimated value, 4d. per lb.	

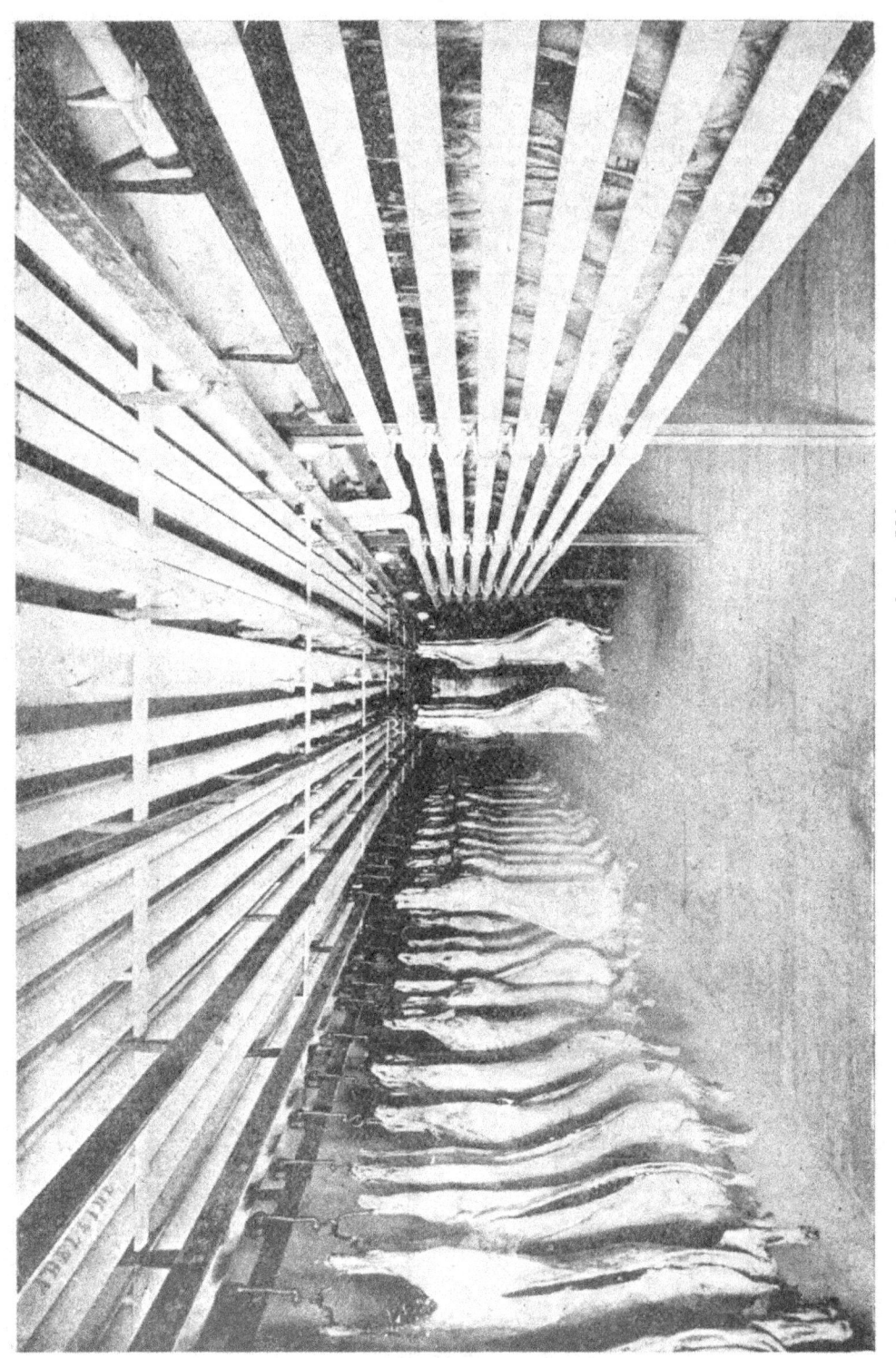

Freezing Chamber, Government Freezing Works, Port Adelaide.

Useful Information on our Land Laws.

Being Extracts taken from Mr. T. Duffield's Book
by his kind permission.

———

Areas suitable for farms may be taken up of sizes varying according to the quality, of a value of £5,000, or of pasture land only for 5,000 sheep, or in dry areas 10,000 sheep. These lands may be held either on Perpetual Lease or on Agreement to Purchase. In the latter case the payments, made half-yearly, go towards purchase money, and on 60 such payments being made the purchase is complete.

Land held under the Pinnaroo Railway Act, which must be held on Agreement to Purchase, may be purchased outright at any time. Completion of purchase of other land, except re-purchased land, may be made after holding it for six years.

If taken on Perpetual Lease the annual rental will be according to the value of the land, from about $\frac{1}{4}$d. to about 6d. per acre. Re-purchased lands must be taken on agreement to purchase, and may be held up to £2,000 worth unimproved value, improved blocks and grazing land up to £4,000 worth. The purchase money must be paid in 70 half-yearly instalments (the first ten payments will be interest only, which will be 4% on the purchase money). Purchase may be completed by paying balance of purchase money after holding the land nine years.

Homestead Block up to £100 worth may be taken on Perpetual Lease or Agreement to Purchase. These are meant for workmen's homes, not to make a living on, but to devote their spare time upon while not employed elsewhere.

Leases are issued for 21 years, 640 acres for removal of guano or other deposits, and for lands resumed for water, or if artesian five square miles, also for small areas for sites for manufactures, wharves, buildings, or other purposes.

Town allotments are small blocks for townships   These are sold at auction for cash.

How to Get the Land.

The Land Board, composed of three members, arranges the subdivision of lands and fixes the price at which block is to be offered.   When approved by the Commissioner, the lands are gazetted as open to application.   A month's notice is usually given, during which applications are made. These must be accompanied by 20 per cent. of the annual rent if perpetual lease is applied for, or a-half year's instalment if desired on a covenant to purchase.   The applicant has a choice whether he will take a lease or an agreement to purchase except for repurchased lands, and lands within the Schedule to the Pinnaroo Railway Act, which are offered on agreement to purchase only.

Agreements and Leases are liable to forfeiture if payments due thereunder are six months in arrear and remain unpaid for three months, after same has been demanded, or for breach of any of the other covenants or conditions.

All land disposed of under Perpetual Lease or Agreement to Purchase will be subject to Land Tax.

Transfers.

Leases and agreements may be transferred after being held for five years, subject to the recommendation of the

Board, and the approval by the Commissioner, but not so as to let any person acquire more land than is allowed to be held as already stated, and all rents or instalments due must be paid except in dry areas where terms can be arranged.

PASTORAL ACT, 1904.

This Act deals with Crown Lands that do not come within the scope of the " Crown Lands Act, 1903."

The Act is administered, under the Commissioner of Crown Lands, by a Board appointed by the Governor, consisting of three members—the Surveyor-General as Chairman, another Civil Servant, and a member selected from outside the Service.   The Board holds office twelve months, the members being eligible for re-appointment.   The Board meets when required.   No member is to be interested in any application dealt with by the Board, and any act done contrary to this restriction is null and void.

The Board's duties are to decide upon the area, rent, and term of lease of land, and to allot same.   Valuations may be made by one member and confirmed by the Board ; evidence of applicants may also be taken by one member.   The Board may require the attendance of applicants and other persons for evidence and may require the production of documents.

In fixing size of blocks, regard is paid to natural features, so as to utilise improvements and waters to each block as equally as possible.

The amount to be paid for any improvement is fixed, distinguishing between amounts payable to the Crown and to the outgoing lessee.   Notice of land available is published in the *Gazette*, showing area, situation, term of rent of each block, price to be paid for improvements, if any, and to whom payable, and cost of valuing such improvements.

Any land not applied for within a month of notice may be re-offered at reduced prices, and so on at intervals of three months until applied for, but if improvements belong to outgoing lessee the price shall not be reduced except for depreciation, until the rent has been reduced 50 per cent., after which rent and price for improvements shall be reduced proportionately after notice to the outgoing lessee of intention to reduce. The outgoing lessee may appeal to "The Tenant's Relief Board" to fix the rent and price for improvements. This Board may fix any price it thinks fit without further appeal.

TRAVELLING STOCK.

Any person before crossing a run with stock, must give to the lessee or person in charge the usual notice in writing, of date of entry and route to be taken. The stock must take the most direct track at the rate of at least five miles a day. If travelling for feed, payment must be made to the lessee at the rate of sixpence per 100 sheep and sixpence per 20 cattle, for every day or part of a day the stock remains on the run.

WATER.

If a lessee discovers artesian water on his run at least ten miles from any other artesian supply on his run that yields not less than 5,000 gallons per day of water suitable for stock, he is entitled to 100 square miles of land surrounding the well, rent free, for ten years, for each well so discovered up to four.

MISCELLANEOUS.

The Commissioner may give permission to any person to erect gates on any roads outside corporative towns and District Councils, and let the right to depasture such roads.

Anyone who injures, destroys, or leaves open such gate is liable to a fine of £50, or six months imprisonment.

VERMIN AND NETTING.

Barb wire and netting of all fences vermin-proofed by a lessee at his own cost remains his property, to be valued and paid for at the termination of the lease by the incoming lessees, the same as all other improvements.

The cost at the nearest port or railway station of barb wire and netting required for vermin proofing boundary fences may be advanced to the lessee by the Commissioner in certain cases, on the recommendation of Board, after wire and netting to the amount of such cost has been utilised in vermin-proofing boundary fences. These advances bear interest at 4½ per cent. per annum, principal and interest being repaid in twenty equal annual instalments of £7 13s. 9d. for every £100 advanced.

If the lease expires before all these instalments are repaid the balance must be paid on the date of such expiry, unless the incoming tenant agrees to pay the same, or the amount may be repaid on six months notice. An incoming tenant is allowed similar credit. The Commissioner may erect such vermin-proof fences, the cost being repayable with interest as described above, provided that a plan showing situation and description of fence and the country to be benefited, names of lessees, and the proportion of cost to be

charged to each, is first laid before Parliament and a copy
sent to all persons proposed to be rated, and afterwards re-
solutions are passed in Parliament approving the erection of
the fence.

No transfer or absolute surrender of lands in a vermin
district is to be allowed until all loans for vermin fences are
repaid, unless the Commissioner is satisfied that payment on
account of the lease will be duly made.   Nor shall any lessee
transfer or surrender his lease while any money is due by him
for wire netting or barbed wire.

Carriage of Live Stock.

RATE PER MILE (S.A.) for cattle van (capacity, 8 averaged sized bullocks) or ordinary sheep van (capacity, say 100 shorn wethers). Minimum charge per van 10s.

Pigs in double tier sheep vans will be charged 50 per cent. additional to sheep van rates. Minimum, 15s.

Pigs and sheep in double tier sheep vans will be charged 25 per cent. additional to sheep van rates. Minimum, 12s. 6d.

Cattle vans do not run on the South or narrow gauge lines.

TABLE OF TRUCKING RATES TO ADELAIDE.

Station.				Miles.		Van Rate. £ s. d.	Cattle Rate. £ s. d.
Aldgate ..	..	..	..	22	..	0 12 10 ..	—
Ambleside	..	..	..	28	..	0 16 4 ..	—
Anama ..	..	..	..	100	..	2 18 4 ..	—
Ararat (Vic.)	..	..	..	353	..	9 11 5 ..	—
Avenue Range ..	..	..	...	270	..	6 7 6 ..	—
Balaklava	..	..	..	67	..	1 19 1 ..	—
Balhannah	..	..	..	29	..	0 16 11 ..	—
Ballarat (Vic.) ..	..	..	..	416	..	11 6 5 ..	—
Barunga Gap	..	..	..	124	..	3 10 4 ..	—
Beachport	..	..	..	354	..	7 15 6 ..	—
Belair ..	..	..	..	14	..	0 10 0 ..	—
Belalie North	..	..	..	164	..	4 9 2 ..	—
Beltana ..	..	..	..	354	..	7 15 6 ..	—
Binnum ..	..	..	..	222	..	5 11 6 ..	—
Blackfellow's Creek	..	..	..	342	..	7 11 6 ..	—
Black Rock	..	..	..	168	..	4 10 10 ..	—
Blackwood	..	..	..	12	..	0 10 0 ..	—
Blyth ..	..	..	..	93	..	2 14 3 ..	—
Bordertown	..	..	..	183	..	4 17 1 ..	—
Bowmans	..	..	..	76	..	2 4 4 ..	—

Station.	Miles.	Van Rate.			Cattle Rate.		
		£	s.	d.	£	s.	d.
Bridgewater	24 ..	0	14	0 ..	—		
Brinkworth	105 ..	3	0	10 ..	—		
Broken Hill (N.S.W.)	335 ..	8	3	5 ..	—		
Bruce	222 ..	5	11	6 ..	—		
Bugle Ranges	40 ..	1	3	4 ..	—		
Burra	102 ..	2	19	4 ..	3	11	2
Burrungule	315 ..	7	2	6 ..	—		
Bute	132 ..	3	14	4 ..	—		
Callington	45 ..	1	6	3 ..	—		
Caltowie	148 ..	4	2	4 ..	—		
Corron	130 ..	3	13	4 ..	—		
Carrieton	199 ..	5	3	9 ..	—		
Cockburn	299 ..	6	17	2 ..	—		
Compton	306 ..	6	19	6 ..	—		
Cooke's Plains	86 ..	2	10	2 ..	—		
Coonalpyn	115 ..	3	5	10 ..	—		
Coward Springs	521 ..	10	11	2 ..	—		
Crystal Brook	152 ..	4	4	2 ..	—		
Currency Creek	67 ..	1	19	1 ..	—		
Custon	197 ..	5	2	11 ..	—		
Dry Creek	7 ..	0	10	0 ..	0	12	0
Ediowie	311 ..	7	1	2 ..	—		
Eudunda	69 ..	2	0	3 ..	2	8	4
Eurelia	190 ..	5	0	0 ..	—		
Farina	409 ..	8	13	10 ..	—		
Farrell's Flat	88 ..	2	11	4 ..	3	1	7
Finniss	61 ..	1	15	7 ..	—		
Frances	217 ..	5	9	10 ..	—		
Freeling	37 ..	1	1	7 ..	1	5	11
Gawler	25 ..	0	14	7 ..	0	17	6
Gemmells	44 ..	1	5	8 ..	—		
Georgetown	130 ..	3	13	4 ..	—		
Gilberts	62 ..	1	16	2 ..	—		
Gladstone	137 ..	3	16	10 ..	—		
Glencoe	306 ..	6	19	6 ..	—		
Glenroy	260 ..	6	4	2 ..	—		
Goolwa	72 ..	2	2	0 ..	—		
Gordon	256 ..	6	2	10 ..	—		
Green's Plains W.	110 ..	3	3	4 ..	—		
Gulnare	121 ..	3	8	10 ..	—		
Halbury	74 ..	2	3	2 ..	—		
Hallett	120 ..	3	8	4 ..	4	2	0
Hamilton (Vic.)	419 ..	11	8	5 ..	—		
Hamley Bridge	45 ..	1	6	3 ..	1	11	6
Hammond	214 ..	5	8	10 ..	—		
Hanson	94 ..	2	14	10 ..	3	5	10
Hawker	276 ..	6	9	6 ..	—		
Hergott Springs	441 ..	9	4	6 ..	—		

Station.				Miles.		Van Rate.				Cattle Rate		
						£	s.	d.		£	s.	d
Hookina	..	..	..	287	..	6	13	2	..		—	
Hoyleton	..	..	..	80	..	2	6	8	..		—	
Huddleston	..	..	..	144	..	4	0	4	..		—	
Hynam ..	..	..	••	234	..	5	15	6	..		—	
Jamestown	..	..	..	156	..	4	5	10	..		—	
Kadina ..	..	..	..	118	..	3	7	4	..		—	
Kalangadoo	..	..	..	286	..	6	12	10	..		—	
Kaniva (Vic.)	..	..	..	211	..	6	2	11	..	1	13	7
Kapunda	..	..	..	48	..	1	8	0	..		—	
Keith ..	..	..	..	155	..	4	5	5	..		—	
Kingston	..	..	..	293	..	6	15	2	..		—	
Kingswood	..	..	..	228	..	5	13	6	..		—	
Koolunga	..	..	..	108	..	3	2	4	..		—	
Kulpara ..	..	..	..	100	..	2	18	4	..		—	
Kybunga	..	..	..	88	..	2	11	4	..		—	
Kybybolite	..	..	..	228	..	5	13	6	..		—	
Laura ..	..	..	..	144	..	4	0	4	..		—	
Lameroo	..	..	..	139	..	3	17	10	..		—	
Leigh's Creek	..	..	..	374	..	8	2	2	..		—	
Littlehampton	..	..	..	34	..	0	19	10	..		—	
Lucindale	..	..	..	262	..	6	4	10	..		—	
Lyndhurst	..	..	..	393	..	8	8	6	..		—	
Mannahill	..	..	..	234	..	5	15	6	..		—	
Moorlands	..	..	..	87	..	2	10	9	..		—	
Manoora	..	..	..	76	..	2	4	4	..	2	13	2
Meadows	..	..	..	324	..	7	5	6	..		—	
Melbourne (Vic.), Newmarket			..	506	..	13	16	5	..		—	
Mern Merna	..	..	..	297	..	6	16	6	..		—	
Middleton	..	..	..	75	..	2	3	9	..		—	
Milang ..	..	..	..	65	..	1	17	11	..		—	
Millicent	..	..	..	333	..	7	8	6	..		—	
Mingary ..	..	..	..	284	..	6	12	2	..		—	
Mintaro ..	..	..	..	83	..	2	5	5	..	2	18	1
Mitcham	..	..	..	6	..	0	10	0	..		—	
Mona ..	..	..	..	133	..	3	14	10	..		—	
Monarto South	..	..	..	52	..	1	10	4	..		—	
Moockra	..	..	..	206	..	5	6	2	..		—	
Moonta ..	..	..	..	135	..	3	15	10	..		—	
Morgan ..	..	..	..	105	..	3	0	10	..	3	13	0
Mount Barker ..	..	..	..	35	..	1	0	5	..		—	
Mount Barker Junction		..	..	32	..	0	18	8	..		—	
Mount Bryan ..	..	..	..	111	..	3	3	10	..	3	16	7
Mount Dutton ..	..	..	..	663	..	12	18	6	..		—	
Mount Gambier	..	..	..	306	..	6	19	6	..		—	
Mount Lofty	..	..	..	20	..	0	11	8	..		—	
Mount Mary	..	..	..	91	..	2	13	1	..	3	3	8
Murray Bridge ..	..	..	..	61	..	1	15	7	..		—	
Nackara ..	..	..	..	183	..	4	17	1	..		—	
Nairne ..	..	..	..	35	..	1	0	5	..		—	

CARRIAGE OF LIVE STOCK.

Station.				Miles.		Van Rate.				Cattle Rate.		
						£	s.	d.		£	s.	d.
Naracoorte	..	..	..	241	..	5	17	10	..	—		
Nhill (Vic.)	..	..	..	236	..	6	12	11	..	—		
Olary	..	..	..	257	..	6	3	2	..	—		
Oodla Wirra	..	..	..	170	..	4	11	8	..	—		
Oodnadatta	..	..	..	688	..	13	6	10	..	—		
Orroroo	..	..	..	176	..	4	14	2	..	—		
Owen	..	..	..	55	..	1	12	1	..	—		
Parachilna	..	..	..	331	..	7	7	10	..	—		
Paratoo	..	..	..	193	..	5	1	3	..	—		
Paskeville	..	..	..	106	..	3	1	4	..	—		
Polly's Well	..	..	..	106	..	3	1	4	..	—		
Penola	..	..	..	271	..	6	7	10	..	—		
Parrakie	..	..	..	123	..	3	9	10	..	—		
Petersburg	..	..	..	155	..	4	5	5	..	—		
Parilla	..	..	..	148	..	4	2	4	..	—		
Pinnaroo	..	..	..	162	..	4	8	4	..	—		
Port Adelaide	..	..	..	8	..	0	10	0	..	0	12	0
Port Augusta	..	..	..	260	..	6	4	2	..	—		
Port Elliot	..	..	..	78	..	2	5	6	..	—		
Port Pirie	..	..	..	169	..	4	11	3	..	—		
Port Wakefield	..	..	..	83	..	2	8	5	..	—		
Quorn	..	..	..	235	..	5	15	10	..	—		
Reedy Creek	..	..	..	281	..	6	11	2	..	—		
Rendelsham	..	..	..	341	..	7	11	2	..	—		
Riverton	..	..	..	63	..	1	16	9	..	2	4	1
Roseworthy	..	..	..	31	..	0	18	1	..	1	1	8
Saddleworth	..	..	..	69	..	2	0	3	..	2	8	4
Saints	..	..	..	72	..	2	2	0	..	—		
Salisbury	..	..	..	13	..	0	10	0	..	0	12	0
Sandergrove	..	..	..	56	..	1	12	8	..	—		
Serviceton	..	..	..	197	..	5	2	11	..	—		
Silverton	..	..	..	318	..	7	12	11	..	—		
Smithfield	..	..	..	19	..	0	11	1	..	0	13	4
Snowtown	..	..	..	118	..	3	7	4	..	—		
South Hummocks	..	..	..	90	..	2	12	6	..	—		
Stewart's Range	..	..	..	248	..	6	0	2	..	—		
Stirling North	..	..	..	255	..	6	2	6	..	—		
Stockport	..	..	..	50	..	1	9	2	..	1	15	0
Stockyard Creek	..	..	..	52	..	1	10	4	..	—		
Strangways Springs	..	..	..	544	..	10	18	10	..	—		
Strathalbyn	..	..	..	51	..	1	9	9	..	—		
Struan	..	..	..	252	..	6	1	6	..	—		
Stuart's Creek	..	..	..	504	..	10	5	6	..	—		
Sutherlands	..	..	..	78	..	2	5	6	..	—		
Sydney, N.S.W.	..	..	..	1,090	..	26	15	0	..	—		
Tailem Bend	..	..	..	76	..	2	4	4	..	—		
Tantanoola	..	..	..	324	..	7	5	6	..	—		
Tarlee	..	..	..	55	..	1	12	1	..	1	18	6

Station.				Miles.		Van Rate. £ s. d.				Cattle Rate. £ s. d.		
Terowie	..	..	..	140	..	3	18	4	..	4	14	0
Tintfnara	..	..	..	132	..	3	14	4	..	—		
Ucolta	..	..	..	163	..	4	8	9	..	—		
Ulooloo	..	..	..	126	..	3	11	4	..	4	5	7
Victor Harbor	..	..	..	81	..	2	7	3	..	—		
Wallaroo	..	..	..	124	..	3	10	4	..	—		
Walloway	..	..	..	183	..	4	17	1	..	—		
Wandilo	..	..	..	297	..	6	16	6	..	—		
Warnertown	..	..	..	161	..	4	7	11	..	—		
Warrina	..	..	..	634	..	12	8	10	..	—		
Wasleys	..	..	..	37	..	1	1	7	..	1	5	11
Willamulka	..	..	..	127	..	3	11	10	..	—		
William Creek	..	..	..	567	..	11	6	6	..	—		
Willochra	..	..	..	246	..	5	19	6	..	—		
Wilson	..	..	..	266	..	6	6	2	..	—		
Winnininnie	..	..	..	216	..	5	9	6	..	—		
Wirrawilla	..	..	..	421	..	8	17	10	..	—		
Wirrega	..	..	..	170	..	4	11	8	..	—		
Woods	..	..	..	57	..	1	13	3	..	—		
Wolseley	..	..	..	192	..	5	0	10	..	—		
Yacka	..	..	..	113	..	3	4	10	..	—		
Yarcowie	..	..	..	134	..	3	15	4	..	4	10	5
Yongala	..	..	..	161	..	4	7	11	..	—		
Yunta	..	..	..	207	..	5	6	6	..	—		

RATES PER HEAD WHEN LESS THAN FULL TRUCK LOADS.

	25 miles.	50 miles.	75 miles.	100 miles.	Each additional 25 miles or part thereof.
Pigs or Sheep ..	6d.	1/-	1/5	1/9	3d.
					Each additional 50 miles or part thereof.
Calves under six months old ..	1/-	2/-	—	3/6	1/-
Cattle and Horses in cattle van ..	2/6	5/-	—	9/-	3/-

The minimum charge for each consignment is one-third of full truck rate, but not less than 5/- will be charged.